Lecture Notes in Mathematics

A collection of informal reports and seminars
Edited by A. Dold, Heidelberg and B. Eckmann, Zürich

134

Larry Smith

University of Virginia, Dept. of Mathematics
Charlottesville, VA/USA

Lectures on the Eilenberg-Moore Spectral Sequence

Springer-Verlag
Berlin · Heidelberg · New York 1970

Title No. 3290

To

Mi – Soo

Introduction

These notes are an outgrowth of lectures that I delivered during the spring of 1969 at Aarhus University. They represent my feeble attempts to organize in a coherent way the circle of ideas revolving about the spectral sequence introduced by Eilenberg and Moore in [18]. The first part of these notes presents a new construction for the spectral sequence based on viewing it as a Kunneth spectral sequence in a suitable category, the category Top/B of spaces over a fixed space B. This idea has also been employed by L. Hodgkin whose work has not yet appeared in print. The category Top/B and its pointed analog offer a suitable setting for many other ideas, constructions and theorems. We reccomend that the interested reader consult [8] [19] [23] [28] [33] and/or [45] for further material in this direction.

The second part of these notes deals with a situation in which the Eilenberg-Moore spectral sequence has proved most tractable. Namely the study of stable Postnikov systems. Most, if not all , of the material of this section is an outgrowth of my joint published [31] and unpublished work with J.C.Moore. I have tried to collect and clarify the results that are spread through [31],[38], [39], and [44] as they apply to a particular problem, namely how does the Pontrjagin ring of a Hopf space depend on the number of k-invariants. These same ideas and techniques have proven useful in other related situations (see for example [37],[40],[41]) and it is hoped they will commend themselves to further study.

The third part of these lectures is concerned with several results that may be obtained from the precursor to the Eilenberg-Moore spectral sequence introduced by J.F. Adams in [1]. Many of the results we discuss in this part are an outgrowth of my unpublished work with Alan Clark. I believe that these results have been known to the experts for some time. They demonstrate the distinct advantage to be obtained form the algebraic approach of [1] and [18] in certain situations.

There are individual introductions to the three separate parts of these notes and we refer to them for a more detailed summary of the material covered.

I would like to express thanks to my many collaborators through the years, P.F. Baum, A. Clark, J.C. Moore and R.E. Stong, for the help and guidance that they have provided me with. I am most grateful to Professor Svend Bundgaard of the Matematisk Institut at Aarhus for his kind invitation to deliver these lectures and to the participants in the topology seminar, who helped me to clarify many obscure points. I am indebted to the Air Force Office of Scientific Research for a Post Doctoral Fellowship and to M. Leon Motchane of I.H.E.S. who provided me with a setting conducive to writting these notes.

Finally I would like to express my gratitude to my wife who typed a large portion of these notes and tolerated my foul moods during their writting and the research that led up to them.

Charlottesville LARRY SMITH

TABLE OF CONTENTS

CHAPTER I

The Eilenberg-Moore Spectral Sequence as a Kunneth SpectralSequence

As noted in the introduction the spectral sequence introduced by Eilenberg-Moore [18] has proven to be a useful computational tool in algebraic topology. We recall that this spectral sequence is defined in the following situation. We suppose that k is a field and denote singular cohomology with coefficients in k by $H^*(\ ;k)$. We suppose given a fibre square

$$\mathfrak{F} \qquad\qquad \begin{array}{ccc} E & \xrightarrow{\ g\ } & E_0 \\ \pi\big\downarrow & & \big\downarrow\pi_0 \\ B & \xrightarrow{\ f\ } & B_0 \end{array}$$

i.e.

(1) $\pi_0 : E_0 \longrightarrow B_0$ is a fibration

(2) $\pi : E \longrightarrow B$ is the fibration induced from π_0 by pullback along f.

Let us assume also that B_0 is 1-connected. Then under suitable finiteness conditions Eilenberg-Moore construct a spectral sequence $\{E_r(\mathfrak{F}), d_r(\mathfrak{F})\}$ such that

$$E_r(\mathfrak{F}) \Longrightarrow H^*(E\ ;k)$$

$$E_2^{p,q}(\mathfrak{F}) = \mathrm{Tor}^{p,q}_{H^*(B_0;k)}(H^*(B;k)\ ,\ H^*(E_0;k)\ .$$

(For a more complete discussion see e.g. [18], [39] or the discussion below.)

Our goal in this chapter is to show how a change in viewpointleads quite naturally to a simple geometric construction of this spectral sequence. Rather than view the spectral sequence as a functor of the fibre square $\mathfrak{F}$ we view it as a functor of the two maps

$$f: B \longrightarrow B_0 \longleftarrow E_0 : \pi_0 .$$

These maps are viewed as topological spaces "parameterized" over B_0. After setting up an appropriate category of such objects we will find that the Eilenberg-Moore spectral sequence may be very profitably viewed as a Kunneth spectral sequence. With this in mind we proceed to apply an idea of Atiyah [6] (see also [4 ;I],[11;§ 8]) to construct the desired spectral sequence.

The geometric nature of the construction that we present yields several bonuses in the forms of extensions of the theory to generalized homology theories and new relations with the Steenrod algebra. These are dealt with in [43]. In the interest of simplicity and clarity we will restrict our discussion to the case of ordinary cohomology with coefficients in a field and refer the reader to [43,I] for the relation to the Steenrod algebra and to [43,II] for extensions to generalized homology theories.

ACKNOWLEDGEMENT: Recently D.L.Rector [34] has given a semi-simplicial "geometric" construction of the Eilenberg-Moore spectral sequence. An approach very similar to our own has also been worked out by A. Heller [19]. My own work (presented below and in [43]) is an outgrowth of my attempts to understand the work of L. Hodgkin in [20]. I am greatly indebted to L.Hodgkin for sending me a copy of [20]. Since working these ideas out for myself I learned that Hodgkin has persued a similar course, although this as yet remains unpublished.

§1. Reformulation: A Generalized Kunneth Theorem

Our objective in this section is to reformulate the situation to which the Eilenberg-Moore spectral sequence applies so as to make more plausable the geometric construction of the spectral sequence we will give subsequently.

We will adopt the point of view first employed by Hodgkin in [20] that the Eilenberg-Moore spectral sequence should be the Kunneth spectral sequence for a suitable cohomology theory on a suitable category. Our first task is to describe the correct category.

We will need the following.

Convention: Throughout these lectures the word space will mean a topological space with compactly generated topology [48] of the homotopy type of a CW - complex. The results of [29] and [48] will assure us that, with the obvious care, our constructions do not carry us outside of this category.

With Hodgkin's idea in mind, let us fix throughout the remainder of this discussion a topological space B. We introduce the category Top/B of topological spaces over B. An object of Top/B is a map $f: T(f) \longrightarrow B$; we will follow the language of bundle theory and think of the domain $T(f)$ of f as the total space of f . If f and g are spaces over B , a morphism $\alpha : f \longrightarrow g$ in Top/B is a commutative triangle

$$\begin{array}{ccc} T(f) & \xrightarrow{T(\alpha)} & T(g) \\ & {}_{f}\searrow \quad \swarrow_{g} & \\ & B & \end{array}$$

of morphisms of spaces.

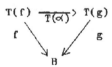

We note that the category Top/B has products; the familar fibre product. Namely if

$$f: T(f) \longrightarrow B \longleftarrow T(g) : g$$

are spaces over B, we set

$$T(f) \underset{B}{\times} T(g) = \{ (x,y) \in T(f) \times T(g) \mid f(x) = g(y) \}$$

and find that we may construct a diagram

$$\begin{array}{ccc} T(f) \underset{B}{\times} T(g) & \xrightarrow{T(\pi_g)} & T(g) \\ {\scriptstyle T(\pi_f)}\downarrow & f \underset{B}{\times} g & \downarrow {\scriptstyle g} \\ T(f) & \xrightarrow[f]{} & B \end{array}$$

where

$$(f \underset{B}{\times} g)(x,y) = f(x) = g(y).$$

The morphisms

$$f \xleftarrow[\pi_f]{} f \underset{B}{\times} g \xrightarrow[\pi_g]{} g$$

are easily seen to be a product of f and g in the category Top/B .

The category Top/B has many of the properties of the category of spaces, and it is this that we will exploit. This idea seems to have occured to several people simultaneously. I learned of it from I.M. James and J-P. Meyer [23] [28], through conversations and lectures. Other sources (there are perhaps more) are Becker and Milgram [8] and McClendon [33].

Suppose that $\mathcal{H}^*$ is an unreduced cohomology theory on the category of spaces. We may prolong $\mathcal{H}^*$ to a sort of cohomology theory on the category Top/B by setting

$$\mathcal{H}^*_B(f) = \mathcal{H}^*(T(f)) ,$$

for any $f \in \text{obj}Top/B$. Now a moments reflection shows that the identity map $1_B : B \longrightarrow B$ is a base point (= terminal object) in the category Top/B . Thus the coefficents of the cohomology theory $\mathcal{H}^*_B$ ought to be $\mathcal{H}^*_B(1_B)$ ($= \mathcal{H}^*(B)$) .

Now let us consider what form the Kunneth theorem ought to take for the functor $\mathcal{H}^*_B$ on the category Top/B . Some rather elementary considerations lead us to expect that under suitable regularity conditions we should have a spectral sequence $\{E_r(f\ ,\ g)\ ,\ d_r(f,g)\}$ with

$$E_r(f,g) \Longrightarrow \mathcal{H}^*_B(f \underset{B}{\times} g)$$

$$E_2(f,g) = \text{Tor}_{\mathcal{H}^*_B}(\mathcal{H}^*_B(f)\ ,\ \mathcal{H}^*_B(g))\ .$$

Rewritting this in terms of our original cohomology theory $\mathcal{H}^*$ we find

$$E_r(f,g) \Longrightarrow \mathcal{H}^*(T(f) \underset{B}{\times} T(g))$$

$$E_2(f,g) = \text{Tor}_{\mathcal{H}^*(B)}(\mathcal{H}^*(T(f)),\ \mathcal{H}^*(T(g)))$$

which is exactly what one would expect of an Eilenberg-Moore type spectral séquence for the cohomology theory $\mathcal{H}^*$.

Thus we find that to obtain a spectral sequence of Eilenberg-Moore type we must really discuss the Kunneth theorem in the category Top/B . This change of viewpoint,from that of the fibre square picture, is really more signifigant than it may seem at first glance. The reason is the strong resemblence between Top/B and spaces (see the next section or [23] , [28] , [8] [33]) . For there exists a method, due to Atiyah [6] (see also [4;I] and [15,§8]) for dealing with the Kunneth problem for generalized homology/cohomology theories. The viewpoint that we adopted therefore suggests that we try also to adopt this method to the case at hand, and this is exactly what we shall do. The ensuing construction being little more complex than that of [4;I] or [15;§8] and has been carried out in detail in [43]. The discussion below will be somewhat simplified by restricting ourselves to the case where $\mathcal{H}^*$ is classical cohomology with field coefficents.

Now a moments reflection on Atiyah's method shows that it is homotopy theoretic in nature. Thus we shall want to do homotopy theory in the category of spaces over B. Homotopy thery invariably requires a basepoint and so we will be forced to work in a suitable category of pointed spaces over B. This category is the subject of the next section.

§ 2. The Category of Pointed Spaces Over B.

As in the previous section B will continue to denote a fixed space and Top/B the category of spaces over B.

Definition: A pointed space over B is a pair (f,s) of maps

$$T(f) \xrightarrow{f} B \xrightarrow{s} T(f)$$

such that

$$fs = 1_B : B \longrightarrow B.$$

More formally, a pointed space over B may be viewed as a pair (f,σ) where f is a space over B and $\sigma : 1_B \longrightarrow f$ is a morphism of spaces over B . In either form we shall refer to σ or s as a basepoint for f. We shall also on occasion refer to σ or s as a cross-section to f .

<u>Definition</u>: If (f,s) and (g,t) are pointed spaces over B then a morphism $\varphi : (f,s) \longrightarrow (g,t)$ of pointed spaces over B is a map $T(\varphi): T(f) \longrightarrow T(g)$ such that the diagrams

$$\begin{array}{ccc} T(f) & \xrightarrow{T(\varphi)} & T(g) \\ & {\scriptstyle f}\searrow \quad \swarrow{\scriptstyle g} & \\ & B & \end{array} \qquad\qquad \begin{array}{ccc} T(f) & \xrightarrow{T(\varphi)} & T(g) \\ & {\scriptstyle s}\searrow \quad \swarrow{\scriptstyle t} & \\ & B & \end{array}$$

commute.

Thus a morphism, $\varphi : (f,\sigma) \longrightarrow (g,\tau)$, of pointed spaces over B may be viewed as a morphism $\varphi : f \longrightarrow g$ of spaces over B that preserves basepoints $\sigma : 1_B \longrightarrow f$ and $\tau : 1_B \longrightarrow g$.

<u>Definition</u>: The category whose objects are the pointed spaces over B, and whose morphisms are morphisms of pointed spaces over B, is denoted by $(Top/B)_*$, and is called the category of pointed spaces over B.

Our objective in the category $(Top/B)_*$ the basic ingredients of homotopy theory. (See e.g. [8] [23] [28] [33].) There are no suprises and we have merely to mimic the usual constructions for $(Top/*)_*$, with a little more attention to basepoints. What little proof is needed is left to the diligent reader as exercises.

<u>Basepoints</u>: As one would expect the category $(Top/B)_*$ is of course pointed. The object $(1_B, 1_B)$ given by

$$1_B : B \longrightarrow 1_B \qquad\qquad 1_B : B \longrightarrow B$$

is both terminal and initial. For if (f,s) is a pointed space over B there are unique morphisms

$$(1_B, 1_B) \longrightarrow (f,s) \longrightarrow (1_B, 1_B)$$

of pointed spaces over B given on total spaces by

$$B \xrightarrow{s} T(f) \xrightarrow{f} B .$$

If (f,s) and (g,t) are pointed spaces over B, then the trivial morphism $*_B : (f,s) \longrightarrow (g,t)$, is by definition the composite

$$*_B : (f,s) \longrightarrow (1_B, 1_B) \longrightarrow (g,t) .$$

Some Terminology: In dealing with pointed spaces over B, we will adopt the terminology common to fibre space theory. Thus if (f,s) is a pointed space over B, we refer to

$$f : T(f) \longrightarrow B$$

as the projection, $T(f)$ as the total space, and

$$s : B \longrightarrow T(f)$$

as the cross-section of f. For any $b \in B$, the fibre of f at B, denoted by $F_b(f)$ is by definition

$$F_b(f) = \{ x \in T(f) \mid f(x) = b \} .$$

If $\varphi : (f,s) \longrightarrow (g,t)$ is a morphism of pointed spaces over B, then we will write

$$F_b(\varphi) : F_b(f) \longrightarrow F_b(g)$$
$$T(\varphi) : T(f) \longrightarrow T(g)$$

for the maps induced by

Some Functors: There are various relations between the categories Top/B, $(Top/B)_*$ $Top_* = (Top/*)_*$ etc., with which we should deal. Note first that $Top/*$ is just the category of spaces, and $(Top/*)_*$ the category of pointed spaces. There are several forgetful functors

$$(Top/B)_* \xrightarrow{F} Top/B \xrightarrow{T} Top = Top/*$$

and their adjoints

$$Top/B)_* \xleftarrow{G} Top/B \xleftarrow{S} Top$$

which are defined by

G: $Tf \rightsquigarrow Tf \quad B$ where

$$\begin{cases} (Gf)(x) = f(x) & \text{if } x \in Tf \\ (Gf)(x) = x & \text{if } x \in B \end{cases}$$

together with the section

$$s_{Gf} : B \longrightarrow Tf \quad B$$

given by the inclusion.

If $B = *$ this is the usual process of adjoining a disjoint basepoint.

S: $X \rightsquigarrow X \times B$ where

$SX : X \times B \longrightarrow B$ is the projection map.

There is also a functor

$$\Gamma : Top_* \longrightarrow (Top/B)_*$$

given by

$T(\Gamma(X,*) = X \times B$ together with

$\Gamma(X,*) : X \times B \longrightarrow B$; the projection, and

$s_{\Gamma(X,*)} : B \longrightarrow X \times B$ by

$b \longrightarrow (*,b)$.

The functor Γ has a coadjoint

$$\Phi : (Top/B)_* \longrightarrow Top_*$$

where

$$\Phi(f,s) = T(f)/s(B).$$

These functors enjoy many useful properties, most of which are rather obvious and consequences of the adjoint relation [24]. We leave them to the reader.

<u>Homotopies</u>: Suppose that

$$\varphi_0, \varphi_1 : (f,s) \longrightarrow (g,t)$$

are two morphisms of pointed spaces over B. A homotopy (in $(Top/B)_*$ of course) from φ_0 to φ_1 is a family of morphisms in $(Top/B)_*$

$$\psi_r : (f,s) \longrightarrow (g,t) \qquad r \in I=[0,1]$$

continuous in r, such that $\varphi_0 = \psi_0, \quad \varphi_1 = \psi_1.$

Thus the homotopies in $(Top/B)_*$ are what are usually called fiber homotopies.

If (f,s) and (g,t) are pointed spaces over B, then we will denote by [(f,s),(g,t)] the set of homotopy classes of morphisms $(f,s) \longrightarrow (g,t)$ of pointed spaces over B.

The usual properties that one expects of homotopy classes are valid in $(Top/B)_*$, in particular homotopy is an equivalence relation and compositions behave well, and we may pass to the associated homotopy category $[(Top/B)_*]$. As usual [(f,s),(g,t)] is a pointed set with distinguished element the homotopy class of the trivial morphism

$$*_B : (f,s) \longrightarrow (g,t).$$

<u>Cofibrations</u>: A morphism $\varphi : (f,s) \longrightarrow (g,t)$ of pointed spaces over B will be called a cofibration (in $(Top/B)_*$) if it has the homotopy extension property in

$(Top/B)_*$, i.e. if whenever we are given a diagram

$$\begin{array}{ccc} (f,s) & \xrightarrow{\ \varphi\ } & (g,t) \\ \downarrow{\scriptstyle \psi_r} & \swarrow{\scriptstyle \hat{\psi}_0} & \\ (h,u) & & \end{array} \qquad r\varepsilon I$$

in $(Top/B)_*$ that commutes when $t=0$, we may find

$$\hat{\psi}_r : \ (g,t) \longrightarrow (h,u) \ , \ r\varepsilon I, \text{ such that}$$

such that

$$\begin{array}{ccc} (f,s) & \longrightarrow & (g,t) \\ \downarrow{\scriptstyle \psi_r} & \swarrow{\scriptstyle \hat{\psi}_r} & \\ (h,u) & & \end{array}$$

is a commutative diagram in $(Top/B)_*$ for all $t\varepsilon I$. As we shall presently see $(Top/B)_*$ has an abundance of cofibrations.

<u>Mapping Cylinders</u> : Suppose that $\varphi : (f,s) \longrightarrow (g,t)$ is a morphism in $(Top/B)_*$. We define the mapping cylinder of φ , denoted by $(M_B(\varphi), \ s_{M_B(\varphi)})$ by

$$T(M_B(\varphi)) = T(g) \cup T(f) \times I \Big/ \begin{array}{l} (x,0) \sim T\varphi(x); \ x\varepsilon T(f) \\ (s(b),r) \sim t(b); \ b\varepsilon B, r\varepsilon I. \end{array}$$

$$M_B(\varphi) : T(M_B(\varphi)) \longrightarrow B \ \Big| \ \begin{array}{l} M_B(\varphi)(y) = g(y); \ y\varepsilon T(g) \\ M_B(\varphi)(x,r) = f(x); \ x\varepsilon T(f), \\ \qquad r\varepsilon I. \end{array}$$

The cross-section

$$s_{M_B(\varphi)} : B \longrightarrow T(M_B(\varphi))$$

is given in the obvious manner by

$$s_{M_B(\varphi)}(b) = t(b) = (s(b), r); \ b\varepsilon B, \ r\varepsilon I.$$

We then have natural morphisms of spaces over B

$$(f,s) \xrightarrow{\varphi'} (M_B(\varphi), s_{M_B(\varphi)}) \xrightarrow{\varphi''} (g,t)$$

given on total spaces by

$$T(\varphi')(x) = (x,1) \, \varepsilon T(M_B(\varphi)); \quad x \varepsilon T(f).$$

$$\begin{cases} T(\varphi'')(y) = y & \text{for } y \varepsilon T(g) \\ T(\varphi'')(x,r) = T(\varphi)(x) & \text{for } x \varepsilon T(f), \; r \varepsilon I. \end{cases}$$

It is readily verified that φ' is a cofibration and φ'' a homotopy equivalence. Thus we find that just as in the category Top_* every map factors into the composition of a cofibration and a homotopy equivalence.

Observe that on fibers $M_B(\ \ )$ is the usual mapping cylinder construction.

Mapping Cones : Suppose that $\varphi : (f,s) \longrightarrow (g,t)$ is a morphism in $(Top/B)_*$. We define the mapping cone of φ, denoted by $(C_B(\varphi), s_{C_B(\varphi)})$ by

$$T(C_B(\varphi)) = T(g) \cup T(f) \times I \Big/ \begin{array}{l} (x,0) \sim T\varphi(x) \quad ; \quad x \varepsilon T(f) \\ (s(b),r) \sim t(b); \; b \varepsilon B, \; r \varepsilon I \\ (x',1) \sim (x'',1) \quad \text{iff} \quad f(x') = f(x'') \\ \qquad \text{for } x', x'' \varepsilon T(f) \end{array}$$

$$C_B(\varphi) : TC_B(\varphi) \longrightarrow B \quad \Big| \quad \begin{array}{l} C_B(\varphi)(y) = g(y); \; y \varepsilon T(g) \\ C_B(\varphi)(x,r) = f(x); \; x \varepsilon T(f), r \varepsilon I. \end{array}$$

The cross-section $s_{C_B(\varphi)}$ is given by

$$s_{C_B(\varphi)}(b) = t(b) = (s(b),r); \; b \varepsilon B, \; r \varepsilon I.$$

There is a natural inclusion

$$(g,t) \xrightarrow{\psi} (C_B(\varphi), s_{C_B(\varphi)})$$

which is of course a cofibration in $(Top/B)_*$.

On fibres the construction $C_B(\ \ )$ quite evidently coincides with the usual mapping cone construction.

Cofibration Sequences : With the aid of the mapping cone construction over B we may introduce the cofibre of a cofibration. More precisely, suppose

$$\varphi : (f,s) \longrightarrow (g,t)$$

is a cofibration. We define the cofibre of φ to be the morphism

$$\psi : (g,t) \longrightarrow (C_B(\varphi),\ s_{C_B(\varphi)})$$

described above. The triple

$$(f,s) \xrightarrow[\varphi]{} (g,t) \xrightarrow[\psi]{} (C_B(\varphi),\ s_{C_B(\varphi)})$$

will then be referred to as a cofibration sequence. More generally, any sequence isomorphic (in the obvious sense) to a sequence of the above type will be called a cofibration sequence in $(Top/B)_*$.

Suspensions : Suppose that (f,s) is a pointed space over B. We form the suspension of (f,s) in $(Top/B)_*$, denoted by $S_B(f,s)$ or $(S_Bf,\ s_{S_Bf})$ by setting

$$TS_B(f) = T(f) \times I \Big/ \begin{matrix} \left.\begin{matrix}(x',0)\sim(x'',0)\\(x',1)\sim(x'',1)\end{matrix}\right\} \text{ iff } f(x')=f(x'') \quad x',x''\varepsilon T(f) \\ (s(b),r')\sim(s(b),r'');\ r',r''\varepsilon I,\ b\varepsilon B. \end{matrix}$$

the base point being given by

$$s_{S_B(f)}(b) = (s(b),r) \ ; \quad b\varepsilon B,\ r\varepsilon I.$$

The projection $S_B(f)$ is of course given by

$$S_B(f)(x,r) = f(x); \quad x\varepsilon T(f),\ r\varepsilon I.$$

If $\varphi : (f,s) \longrightarrow (g,t)$ is a morphism of pointed spaces over B, then we define the suspension of φ, denoted by $S_B(\varphi)$, by

$$TS_B(\varphi)(x,r) = (T\varphi(x),r), \ x \in T(f), \ r \in I.$$

One readily verifies that

$$S_B(\varphi) : (S_B f, s_{S_B(f)}) \longrightarrow (S_B(g), s_{S_B(g)})$$

is a morphism of pointed spaces over B.

(As we are using Greek letters for morphisms in $(Top/B)_*$ and latin letters for projections, e.g. $f:T(f) \longrightarrow B$, there should be no confusion between the projection $S_B(f)$ and the morphism of pointed spaces over B, $S_B(\varphi)$.)

Indeed $S_B: (Top/B)_* \longrightarrow (Top/B)_*$ is a functor.

As in the category Top_* we may introduce track addition into $[(S_B f, s_{S_B f}), (g,t)]$ and obtain a group structure, which on $[(S_B^2 f, s_{S_B^2 f}), (g,t)]$ is abelian. Moreover, if

$$(f',s') \xrightarrow[\varphi']{} (f,s) \xrightarrow[\varphi'']{} f'',s'')$$

is a cofibration sequence in $(Top/B)_*$, then the sequence

$$[(f',s'),(g,t)] \longleftarrow [(f,s),(g,t)] \longleftarrow [(f'',s''),(g,t)]$$

is exact in the sense appropriate to the number of suspensions present.

Note that if (f,s) is a pointed space over B then the cross-section s provides a base point $s(b) \in F_b(f)$ in each fibre $F_b(f)$ of f. It is easy to see that the fibre of $(S_B f, S_B s)$ is the reduced suspension of the pointed space $(F_b(f), s(b))$.

Puppe Sequences: We begin by supposing that $\varphi : (f,s) \longrightarrow (g,t)$ is a morphism in $(Top/B)_*$. As we have seen from the point of view of homotopy theory in $(Top/B)_*$ there is no loss in generality if we assume φ to be an inclusion; and we will suppose this to be the case. We then have the cofibration sequence

$$(f,s) \xrightarrow{\varphi} (g,t) \xrightarrow{\psi} (C_B(\varphi), s_{C_B}(\varphi))$$

which we may continue to a cofibration sequence

$$(g,t) \xrightarrow{\psi} (C_B(\varphi), s_{C_B}(\varphi)) \xrightarrow{\Delta} (C_B(\psi), s_{C_B}(\psi)).$$

The basic ingrediant in forming the Puppe sequence in $(Top/B)_*$ is the observation that $(C_B(\psi), s_{C_B}(\psi))$ and $(S_B(f), s_{S_B}(f))$ are homotopy equivalent in $(Top/B)_*$. We moreover have a commutative diagram

$$\begin{array}{ccc}
(C_B(\psi), s_{C_B}(\psi)) & \xrightarrow{\varsigma} & (C_B(\Delta), s_{C_B}(\Delta)) \\
\downarrow \cong & & \downarrow \cong \\
(S_B(f), S_B s) & \xrightarrow{S_B(\varphi)} & (S_B(g), S_B t)
\end{array}$$

where ς is the natural map. Thus we may form the infinite puppe suquence

$$(f,s) \xrightarrow{\varphi} (g,t) \xrightarrow{\psi} (C_B(\varphi), s_{C_B}(\varphi)) \xrightarrow{\Delta} (S_B f, S_B s) \xrightarrow{S_B(\varphi)} (S_B(g), S_B(t)) \longrightarrow \cdots$$

By standard arguments we may show that the sequence

$$[(f,s),(h,u)] \xleftarrow{\varphi^*} [(g,t),(h,u)] \xleftarrow{\psi^*} [(C_B(\varphi), s_{C_B}(\varphi)), (h,u)]$$

$$\xleftarrow{\Delta^*} [(S_B f, S_B s),(h,u)] \longleftarrow \cdots$$

is exact in the usual sense. Thus the functor $[\quad ,(h,u)]$ is half exact in the sense of [16].

<u>Products</u>: The category $(Top/B)_*$ has products which are given in a perfectly reasonable fashion by the fibre product construction. Namely, if (f,s) and (g,t) are pointed spaces over B, then their product, denoted by $(f,s) \underset{B}{\times} (g,t)$ or $(f \underset{B}{\times} g, s \underset{B}{\times} t)$ is defined by

$$T(f \underset{B}{\times} g) = T(f) \underset{B}{\times} T(g) = \{(x,y) \in Tf \times Tg \mid f(x) = g(y)\}.$$

The maps being given by

$$f \underset{B}{\times} g : T(f \underset{B}{\times} g) \longrightarrow B \mid f(x) = (f \underset{B}{\times} g)(x,y) = g(y);$$

$$\begin{array}{ccc} & & T(f) \\ & \overset{T(\pi_f)}{\nearrow} & \\ T(f \underset{B}{\times} g) & & \\ & \underset{T(\pi_g)}{\searrow} & \\ & & T(g) \end{array} \qquad \begin{array}{l} T(\pi_f)(x,y) = x \\ x \in Tf, \ y \in Tg \\ T(\pi_g)(x,y) = y \end{array}$$

and we have the familier fibre square diagram

$$\begin{array}{ccc} Tf \underset{B}{\times} Tg & \xrightarrow{T\pi_g} & Tg \\ {\scriptstyle T\pi_f}\downarrow & \searrow^{f \underset{B}{\times} g} & \downarrow{\scriptstyle g} \\ Tf & \longrightarrow & B \end{array}$$

The cross-section $s \underset{B}{\times} t$ is defined by

$$(s \underset{B}{\times} t)(b) = (\, s(b), t(b) \,) \in T(f \underset{B}{\times} g); \ b \in B.$$

It is elementary to verify

$$(f,s) \xleftarrow[\pi_f]{} (f \underset{B}{\times} g, s \underset{B}{\times} t) \xrightarrow[\pi_g]{} (g,t)$$

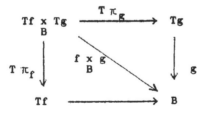

is the product, in the category $(Top/B)_*$, of the objects (f,s) and (g,t).

<u>Smash-Products</u> : The smash product construction may also be adopted to $(Top/B)_*$ in the obvious manner. Precisely, if (f,s) and (g,t) are pointed spaces over B, then their smash product, denoted by $(f,s)\wedge_B(g,t)$ or $(f\wedge_B g,\ s\wedge_B t)$ is defined by

$$T(f\wedge_B g) = Tf\wedge_B Tg = T(f\times_B g) \Big/ \begin{array}{l}(x,s(b))\sim(t(b),y)\\ \text{for } (x,y)\in T(f\times_B g),\\ b\in B\end{array}$$

The map $f\wedge_B g$ is given by

$$f\wedge_B g:\ T(f\wedge_B g)\longrightarrow B \mid (f\wedge_B g)(x\wedge_B y) = t(x) = g(y).$$

The structure of the smash product is completed by defining

$$s\wedge_B t:\ B\longrightarrow T(f\wedge_B g) \mid (s\wedge_B t)(b) = s(b)\wedge_B t(b);\ b\in B,$$

as cross-section. One finds that smash products behave much as expected. For example if we denote by $(S^n{}_B,\ s_{S^n_B})$ the pointed space $\Gamma(S^n)$ over B then

$$(S^n{}_Bf,\ s^n_B s) = (\ S^n_B\wedge_B f,\ S^n_B\wedge_B s)$$

for any pointed space (f,s) over B.

The pointed spaces $(S^n_B,\ s_{S^n_B})$ over B play the role of the n-sphere in the usual category of pointed spaces. We will refer to $(S^n_B,\ s_{S^n_B})$ as the n-sphere over B.

Finally, to complete what we shall need of the elementary

properties of $(Top/B)_*$ we introduce a full subcategory of objects satisfying a local smoothness condition.

<u>Fibrations over B</u>: We will denote by Fib/B the full subcategory of Top/B whose objects are fibrations $f:T(f) \to B$. Similarly we denote by $(Fib/B)_*$ the full subcategory of $(Top/B)_*$ whose objects are pointed spaces (f,s) where $f:Tf \longrightarrow B$ is a fibration. The map $s: B \longrightarrow Tf$ is then a cross-section in the classical sense.

<u>Warning</u>: A fibration over B is to be distinguished from a fibration in the category $(Top/B)_*$.

Notice that we have imposed no connectivity conditions so the functor $G: Top/B \longrightarrow (Top/B)_*$ induces $G:Fib/B \longrightarrow (Fib/B)_*$. It is also important to note (although tedious to verify) that the constructions described above preserve the fibration property.

The fibration condition plays the role of a regularity condition, and will be used as such below in dealing with the Künneth theorem in $(Top/B)_*$.

<u>§3. Cohomology Theories on $(Top/B)_*$</u>

Having provided the category of pointed spaces over B with a fairy rich homotopy structure we turn now to a discussion of homology theories on $(Top/B)_*$. We will adhere to the conventions and notations detailed above, and B continues to denote a fixed space. We will adopt a fairly general definition for homology theories on $(Top/B)_*$, imposing only those conditions required to state the Künneth theorem. Of course if we try to establish this statement we will be forced to impose more stringent conditions.

<u>Definitions</u>: A cohomology theory on $(Top/B)_*$ with values in an abelian category $\mathcal{A}$, consists of

(1) a pointed cofunctor

$$H^* : [(Top/B)_*] \longrightarrow \mathcal{A}_{\mathbb{Z}},$$

where $\mathcal{A}_{\mathbb{Z}}$ denotes the category of $\mathbb{Z}$-graded objects over $\mathcal{A}$;

(2) for each cofibration sequence

$$(f',s') \xrightarrow[\varphi']{} (f,s) \xrightarrow[\varphi'']{} (f'',s'')$$

in $(Top/B)_*$, a morphism of degree 1

$$\Delta(\varphi', \varphi'') : H^*(f',s') \longrightarrow H^*(f'',s'')$$

such that

(3) $\Delta(\varphi', \varphi'')$ is natural, and

(4) for each cofibration

$$(f',s') \xrightarrow[\varphi']{} (f,s) \xrightarrow[\varphi'']{} (f'',s'')$$

in $(Top/B)_*$, the triangle

$$\begin{array}{ccc} H^*(f',s') & \xrightarrow{\quad H^*(\varphi') \quad} & H^*(f'',s'') \\ & {\scriptstyle \Delta(\varphi',\varphi'')} \nwarrow \quad \swarrow {\scriptstyle H^*(\varphi'')} & \\ & H^*(f'',s'') & \end{array}$$

is exact.

There is of course no overwhelming reason to restrict ourselves to $\mathbb{Z}$-graded homology theories, but this restriction will allow a certain simplification in the indexing of our spectral sequence. Notice that when B=point our definition reduces to the usual requirements of a reduced generalized cohomology theory.

The following proposition provides us with the means of defining a large number of useful cohomology theories on $(Top/B)_*$.

Proposition 3.1: If

$$(f',s') \xrightarrow[\varphi']{} (f,s) \xrightarrow[\varphi'']{} (f'',s'')$$

is a cofibration in $(Top/B)_*$, then

$$Tf'/ s'B \longrightarrow Tf/ sB \longrightarrow Tf''/ s''B$$

is a cofibration in Top_* . Thus the functor

$$\Phi : (Top/B)_* \longrightarrow Top_*$$

preserves cofibrations.

Proof: Clearly it suffices to consider the case where

$$(f'',s'') = (C_B(\varphi'), s_{C_B}(\varphi')).$$

Suppose that

$$\psi : (g',t') \longrightarrow (g'',t'')$$

is a morphism of pointed space over B. Then there is induced a morphism

$$\psi/t : Tg'/t'B \longrightarrow Tg''/t''B.$$

One readily checks that

$$TC_B\psi/s_{C_B\psi} = C_*(\psi/t).$$

The result is now immediate. □

Corollary 3.2: Suppose that $\{\mathcal{H}_* \ \Delta^*(,)\}$ is an $\mathcal{A}$-valued homology theory on the category Top_* . Then the composite functor

$$\mathcal{H}^*_B : [(Top/B)_*] \xrightarrow[\Phi]{} Top_* \xrightarrow[\mathcal{H}^*]{} \mathcal{A}_{\mathbb{Z}}$$

is a cohomology theory on $(Top/B)_*$. □

Of particular interest to us will be the cohomology theory $H_B^*(\ ;\pi)$, where π is an abelian group. Recall the n-sphere over B is defined by

$$(S_B^n , s_B^n) = \Gamma (S^n, *) .$$

Following the usual conventions we find that the coefficents of $H_B^*(\ ;\pi)$ are none other than

$$H_B^*(S_B^0 , s_B^0 ;\pi) = H^*(B ;\pi).$$

Note that cohomology theories on $(Top/B)_*$ enjoy the expected properties. For example for future use we note the following stability result whose proof we leave to the reader.

<u>Proposition 3.3</u>: Let $\{ H^* ,\Delta (,)\}$ be a cohomology theory on $(Top/B)_*$. Then for any $(f,s) \in obj(Top/B)_*$ we have a natural isomorphism

$$s_B: H^*(f,s) \longrightarrow H^*(S_B f , S_B s)$$

of degree + 1 . □

With these preliminaries now taken care of we are prepared to investigate the Kunneth theorem in $(Top/B)_*$. As in the case of ordinary spaces the Kunneth theorem ought to concern itself with the following problem:

> Suppose that $H^* : (Top/B)_* \longrightarrow \mathcal{A}_{\mathbb{Z}}$ is an $\mathcal{A}$-valued cohomology theory on the category $(Top/B)_*$.
> Suppose that
>
> $$(f,s) , (g,t) \in obj(Top/B)_* .$$
>
> How does one compute
>
> $$H^*(f \wedge_B g, s \wedge_B t)$$

from a knowledge of

$$H^*(f,s) \text{ and } H^*(g,t) .$$

In order to describe the expected form of the Kunneth theorem we will simplify matters by assuming that the range category of our cohomology theory is the category of modules over some commutative ring R, denoted by $\mathcal{M}/R$. The category of $\mathbb{Z}$ - graded R-modules being denoted by $\mathcal{M}^*/R$. We will also assume that our cohomology theory is multiplicative in the sense of the following definition.

<u>Definition</u>: A cohomology theory

$$\mathbb{H}^* : [(\mathrm{Top}/B)_*] \longrightarrow \mathcal{M}^*/R$$

is said to be multiplicative iff for every pair

$$(f,s)\ ,\ (g,t) \in \mathrm{obj}(\mathrm{Top}/B)_*$$

there is given a natural exterior product

$$\mathbb{H}^*(f,s) \otimes_R \mathbb{H}^*(g,t) \longrightarrow \mathbb{H}^*(f \underset{B}{\wedge} g,\ s \underset{B}{\wedge} t)$$

As usual in the case $(f,s) = (g,t)$ we obtain an R-algebra structure on $\mathbb{H}^*(f,s)$ by composing the exterior product with the map induced by the diagonal. Moreover $\mathbb{H}^*(f,s)$ is a module over $\mathbb{H}^*(S_B^0\ ,\ s_B^0)$ via the cross-product identification

$$(f \underset{B}{\wedge} S_B^0\ ,\ s \underset{B}{\wedge} s_B^0) = (f,s)\ .$$

Thus the following statement makes sense.

<u>Spectral Kunneth Theorem Over B</u>: Suppose that $\{\mathbb{H}^*(\ ), \Delta(\ ,\ )\}$ is a multiplicative cohomology theory on the category $(\mathrm{Top}/B)_*$ with values in $\mathcal{M}^*/R$. If $\mathbb{H}^*(\ )$ is suitably nice then for every pair $(f,s), (g,t) \in \mathrm{obj}(\mathrm{Top}/B)_*$ satisfying suitable regularity conditions, there is a natural cohomology spectral sequence $\{E_r((f,s),(g,t)),\ d_r((f,s),(g,t))\}$ with

$$E_r((f,s),(g,t)) \Longrightarrow H^*(f \underset{B}{\wedge} g,\ s \underset{B}{\wedge} t)$$

$$E_2^{p,q}((f,s),(g,t)) = \mathrm{Tor}^{p,q}_{H^*(S^0_B,s^0_B)}(H^*(f,s),\ H^*(g,t)).$$

The precise restrictions needed to apply Atiyah's method [6] (see also [4] , [15]) can be axiomatized, and this is done in [43]. However in the interest of what we hope is clarity and simplicity we will present the procedure in the next two sections for $H^*_B(\ ;k)$ where k is a field and refer the reader to [43] for the details of the more general case.

§4. The Spectral Kunneth Theorem on (Top/B)$_*$ for Classical Cohomology Theories. The Machinary.

Our objective in this section is to describe the machinary necessary for the construction of the spectral Kunneth theorem for the cohomology theories

$$H^*_B(\ ;k) : [(\mathrm{Top}/B)_*] \longrightarrow {}^*/k$$

when k is a field and B is 1-connected. The construction of the spectral sequence will be given in the next section.

Notations and Conventions: Throughout this section and the next B will denote a fixed space and k a fixed field. The extended cohomology theory

$$H^*_B(\ ;k) : [(\mathrm{Top}/B)_*] \longrightarrow {}^*/k$$

will denoted simply by $H^*_B(\ )$. For B = * we have the usual reduced cohomology with coefficents in k and we write merely $H^*(\ )$ in this case.

The following elementary facts will be essential to our study.

Proposition 4.1: With the above notations

$$H_B^*(\ ) : [(Top/B)_*] \longrightarrow \mathcal{M}^*/k$$

is a multiplicative cohomology theory in the sense of the previous section.

Proof: Merely recall that there is a well defined product

$$H^*(X,A;k) \otimes_k H^*(Y,B;k) \longrightarrow H^*(X \times Y, A \times B ; k)$$

and the result follows from the definitions. □

The construction of the desired Kunneth spectral sequence depends on the ideas embodied in the following definition.

Definition: Let $(f,s) \in obj(Top/B)_*$. An H_B^* - display of (f,s) consists of a sequence of cofibration sequences

$$(f,s) \xrightarrow{\alpha_0} (h_0,u_0) \xrightarrow{\beta_{-1}} (f_{-1},s_{-1})$$
$$(f_{-1},s_{-1}) \xrightarrow{\alpha_{-1}} (h_{-1},u_{-1}) \xrightarrow{\beta_{-2}} (f_{-2},s_{-2})$$
$$\vdots \qquad\qquad \vdots \qquad\qquad \vdots$$
$$(f_{-n},s_{-n}) \xrightarrow{\alpha_{-n}} (h_{-n},u_{-n}) \xrightarrow{\beta_{-n-1}} (f_{-n-1},s_{-n-1})$$
$$\vdots \qquad\qquad \vdots \qquad\qquad \vdots$$

such that

(1) $H_B^*(h_{-i},s_{-i})$ is a projective $H_B^*(S_B^0 ,s_B^0)$ - module for $i = 0,1,2,\ldots$ and

(2) $H_B^*(\alpha_{-i}) : H_B^*(h_{-i},u_{-i}) \longrightarrow H_B^*(f_{-i}, s_{-i})$ is an epimorphism (where $(f_0,s_0) = (f,s)$) for $i = 0 , 1 ,2,\ldots$

Suppose that (f,s) is apointed spave over B and that

$$(f,s) \longrightarrow (h_0,u_0) \longrightarrow (f_{-1},s_{-1})$$
$$\vdots \qquad\qquad \vdots \qquad\qquad \vdots$$
$$(f_{-n},s_{-n}) \longrightarrow (h_{-n},u_{-n}) \longrightarrow (f_{-n-1},s_{-n-1})$$
$$\vdots \qquad\qquad \vdots \qquad\qquad \vdots$$

is an H_B^* - display of (f,s). Then for each fixed integer m our Puppe sequence construction in $(\mathrm{Top}/B)_*$ provides us with maps

$$\Delta_m : (f_{-m-1}, s_{-m-1}) \longrightarrow (S_B f_{-m}, S_b s_{-m}) \quad .$$

Thus by fixing an integer n and forming telescoping mapping cylinders we may obtain the pointed filtered space over B

$$(f_{-n}, s_{-n}) \xrightarrow[\Delta_{n-1}]{} (S_B f_{-n+1}, S_B s_{-n+1}) \hookrightarrow \cdots \xrightarrow[S_B^n \Delta_0]{} (S_B^n f, S_B^n s)$$

Notation: If (g',t') and (g,t) are pointed spaces over B and $\varphi : (g',t') \longrightarrow (g,t)$ is a cofibration, then we will write

$$H_B^*(g,g';t)$$

for

$$H_B^*(C_B(\varphi), s_{C_B(\varphi)}) \ .$$

This notation may be extended to triples in the obvious way and we observe that the usual exact sequences of pairs and triples hold for H_B^* - cohomology.

Our main in H_B^* - displays stems from:

Proposition 4.2: With the above notation, the sequence

$$0 \longleftarrow H_B^*(S_B^n f, S_B^n s) \longleftarrow H_B^*(S_B^n f, S_B^{n-1}; S_B^n s) \longleftarrow \cdots$$

$$\cdots \longleftarrow H_B^*(S_B^{n-i} f_{-i}, S_B^{n-i-1} f_{-i-1}; S_B^{n-i} s) \longleftarrow \cdots$$

$$\cdots \longleftarrow H_B^*(S_B f_{-n+1}, f_{-n}; S_B s_{-n+1})$$

is exact and each of the modules

$$H_B^*(S_B^{n-i} f_{-i} , S_B^{n-i+1} f_{-i+1} ; S_B^{n-i} s_{-i})$$

is a projective $H_B^*(S_B^0, s_B^0)$ - module.

Proof: From the definition of an H_B^* - display it follows that the sequences

$$0 \longleftarrow H_B^*(f_{-i},s_{-i}) \longleftarrow H_B^*(h_{-i},u_{-i}) \longleftarrow H_B^*(f_{-i-1},s_{-i-1}) \longleftarrow 0$$

are exact for $i = 0,1,\ldots,n-1$. Thus by the stability of $H_B^*(\ )$ under suspension (3.3) and the Puppe sequence sequence construction we obtain by the usual splicing argument the long exact sequence

$$0 \longleftarrow H_B^*(S_B^n f, S_B^n s) \longleftarrow H_B^*(S_B^n h_0, S_B^n u_0) \longleftarrow \cdots$$

$$\cdots \longleftarrow H_B^*(S_B^{n-i}\ h_{-i},\ S_B^{n-i} u_{-i}) \longleftarrow \cdots$$

$$\cdots \longleftarrow H_B^*(S_B h_{-n},\ S_B u_{-n}) \quad .$$

Identifying

$$H_B^*\ (S_B^{n-i} h_{-i},\ S_B^{n-i} u_{-i})$$

with

$$H_B^*(S_B^{n-i-1}\ f_{-i-1},\ S_B^{n-i} f_{-i}\ ; S_B^{n-i-1}\ s_{-i-1})$$

we obtain the required exact sequence. By stability and hypothesis

$$H_B^*(S_B^{n-i} h_{-i}, S_B^{n-i} u_{-i}) \cong s^{n-i} H_B^*(h_{-i},u_{-i})$$

is a projective $H_B^*(S_B^0,\ s_B^0)$ - module and the result thus follows via the above identifications. □

Thus for each integer n the filtered pointed space over B

$$(f_{-n},s_{-n}) \hookrightarrow (S_B f_{-n+1}, S_B s_{-n+1}) \hookrightarrow \cdots \hookrightarrow (S_B^n f, S_B^n s)$$

constitutes a geometric analog of a partial projective resolution of $s^n H_B^*(f,s)$ of length n. By increasing n we obtain a more faithful approximation to a resolution over $H_B^*(S_B^0, s_B^0)$ of (f,s) by pointed spaces over B.

Before carrying the study of displays further let us dispose of the existance question for such gadgets.

<u>Proposition 4.3</u>: Let B, k, and H_B^* be as above. Then for any $(f,s) \varepsilon \text{obj}(\text{Top}/B)_*$ there exists an H_B^* display of (f,s).

This is an easy consequence of the following "representation" property of H_B^* which is at the heart of Atiyah's method for constructing Kunneth type theorems (see for example [4] [15]).

<u>Proposition 4.4</u>: Let B , k , H_B^* be as above. Suppose $(f,s) \varepsilon \text{obj}(\text{Top}/B)_*$. Then there exists $(h,u) \varepsilon \text{obj}(\text{Top}/B)_*$ and a morphism

$$\alpha : (f,s) \longrightarrow (h,u)$$

such that

(1) $H_B^*(\alpha) : H_B^*(h,u) \longrightarrow H_B^*(f,s)$ is epic, and

(2) $H_B^*(h,u)$ is a projective $H_B^*(S_B^0 , s_B^0)$ - module.

<u>Proof</u>: Consider the natural mapping

$$a : Tf \longrightarrow (Tf/sB) \times B$$

defined bv

$$a(x) = (q(x), f(x)) \quad : \quad x \varepsilon Tf$$

where $q : Tf \longrightarrow Tf/sB$ is the natural quotient mapping. Let us intruduce the pointed space over B (h,u) given by

$$Th = (Tf/sB) \times B \xrightarrow[\text{projection}]{} B \xrightarrow[b \longrightarrow (sB/sB,b)]{} (Tf/sB) \times B \ .$$

One readily checks that the diagram

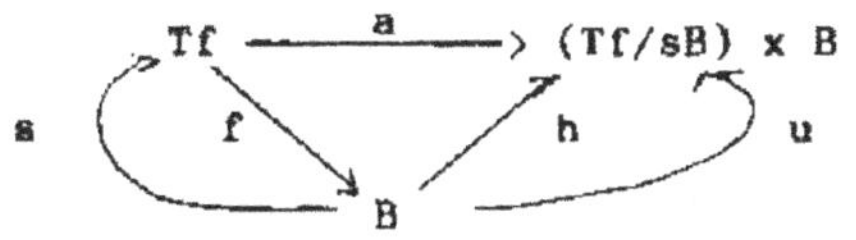

commutes and so a defines a morphism

$$\alpha : (f,s) \longrightarrow (h,u)$$

of pointed spaces over B. Let us see that this morphism has the required properties. Let us first examine

$$H_B^*(\alpha) : H_B^*(h,u) \longrightarrow H_B^*(f,s) .$$

According to our definitions this is the same as the map

$$H^*(\bar{a}) : H^*((Tf/sB) \times B/uB) \longrightarrow H^*(Tf/sB) .$$

Now one readily checks that the map

$$\bar{b} : (Tf/sB) \times B/uB \longrightarrow Tf/sB$$

given by $\bar{b}(x,b) = x$ for $x \in Tf/sB$, $b \in B$, is well defined and

$$\bar{b}\bar{a} = 1 : Tf/sB \longrightarrow Tf/sB .$$

Hence $\bar{a}$ is a coretraction so $H^*(\bar{a})$ is certainly epic.

Next we note that by definition

$$H_B^*(h,u) = H^*((Tf/sB) \times B , uB) = H^*(Tf,sB) \times H^*(B)$$

by the Kunneth theorem for $H^*(\ )$ (recall that k is a field) . Thus $H_B^*(h,u)$ is actually a free $H^*(B) = H_B^*(S_B^0 , s_B^0)$ - module. (This is not suprising, see for example [15;3.2].)

This establishes that α has the required property. □

Proof of 4.3: We apply 4.4 inductively to obtain

$$\begin{array}{ccccc} (f,s) & \longrightarrow & (h_0,u_0) & \longrightarrow & (f_{-1}, s_{-1}) \\ \vdots & & \vdots & & \vdots \\ (f_{-n+1}, s_{-n+1}) & \longrightarrow & (h_{-n+1}, u_{-n+1}) & \longrightarrow & (f_{-n},s_{-n}) \end{array}$$

where

(1) $H^*(\alpha_{-i})$ are epimorphisms for $i = 0, 1, \dots , n-1$; and

(2) $H^*(h_{-i},u_{-i})$ are projective $H_B^*(S_B^0 , s_B^0)$ - modules for $i = 0,\dots,n-1$.

Application of 4.4 to (f_{-n},s_{-n}) then provides the inductive step. □

We turn now to the task of comparing two different H_B^* - dispdays by a morphism.

Proposition 4.5: Let B , k , and H_B^* be as above. Suppose given the following data.

(1) (f,s) , $(g,t) \varepsilon$ obj(Top/B).

(2) $\varphi : (f.s) \longrightarrow (g,t)$ a morphism in (Top/B).

(3) $(g,t) \longrightarrow (k_0,v_0) \longrightarrow (g_{-1},t_{-1}) \longrightarrow (k_{-1},v_{-1}) \longrightarrow \cdots$ an H_B^* - display.

Then there exists an H_B^* - display

$$(f,s) \longrightarrow (h_0,u_0) \longrightarrow (f_{-1},s_{-1}) \longrightarrow \cdots$$

and morphisms making the following diagram

$$\begin{array}{ccccccc} (f,s) & \longrightarrow & (h_0,u_0) & \longrightarrow & (f_{-1},s_{-1}) & \longrightarrow & \cdots \\ \downarrow\varphi & & \downarrow & & \downarrow & & \\ (g,t) & \longrightarrow & (k_0,v_0) & \longrightarrow & (g_{-1},t_{-1}) & \longrightarrow & \cdots \end{array}$$

commute.

This results from an easy induction argument based on the following.

Proposition 4.6: Let B , k , H_B^* be as above. Suppose given a diagram in (Top/B).

$$\begin{array}{ccc} (f,s) & & \\ \varphi\downarrow & & \\ (g,t) & \xrightarrow{\lambda} & (k,v) \end{array}$$

where

(1) $H_B^*(\lambda) : H_B^*(k,v) \longrightarrow H_B^*(g,t)$ is epic ; and

(2) $H_B^*(k,v)$ is a projective $H_B^*(S_B^0 , s_B^0)$ -module.

Then we may complete this to a diagram

$$\begin{array}{ccc} (f,s) & \xrightarrow{\alpha} & (h,u) \\ \varphi\downarrow & & \downarrow\psi \\ (g,t) & \xrightarrow{\lambda} & (k,v) \end{array}$$

wherein

(a) $H_B^*(\alpha) : H_B^*(h,u) \longrightarrow H_B^*(f,s)$ is epic, and

(b) $H_B^*(h,u)$ is a projective $H_B^*(S_B^0, s_B^0)$ - module.

<u>Proof</u>: Consider the map

$$a : Tf \longrightarrow (Tf/sB) \times Tk$$

given by

$$a(x) = (q(x), T(\lambda\varphi)(x)) \quad : \quad x \in Tf \ .$$

Consider the pointed space over B , (h,u) defined by

$$\begin{array}{ccc} Th = (Tf/sB) \times Tk & \xrightarrow{\text{project}} & Tk \\ h \searrow & & \swarrow k \\ & B & \end{array} \quad : \quad u(b) = (sB/s/B, v(b)).$$

One readily checks that a together with projection define morphisms of pointed spaces over B

$$\begin{array}{ccc} (f,s) & \xrightarrow{\alpha} & (h,u) \\ & & \downarrow \psi \\ & & (k,v) \end{array}$$

with the required properties. □

<u>Proof of 4.5</u>: Iterate 4.6. □

<u>Corollary 4.7</u> : Let B , k and H_B^* be as above. Suppose given the following data.

(1) (f,s) , $(g,t) \in obj(Top/B)$,

(2) $\varphi : (f,s) \longrightarrow (g,t) \in morph(Top/B)$,

(3) $(f,s) \longrightarrow (h_0,u_0) \longrightarrow (f_{-1}, s_{-1}) \longrightarrow \ldots$

and

$$(g,t) \longrightarrow (k_0,v_0) \longrightarrow (g_{-1}, t_{-1}) \longrightarrow \ldots$$

H_B^*- displays.

Then there exists an H_B^* - display

$$(f,s) \longrightarrow (h_0', u_0') \longrightarrow (f_{-1}', s_{-1}') \longrightarrow \cdots$$

and morphisms of pointed spaces over B making the diagram

$$(f,s) \longrightarrow (h_0, u_0) \longrightarrow (f_{-1}, s_{-1}) \longrightarrow \cdots$$

$$(f,s) \longrightarrow (h_0', u_0') \longrightarrow (f_{-1}', s_{-1}') \longrightarrow \cdots$$

$$(g,t) \longrightarrow (k_0, v_0) \longrightarrow (g_{-1}, t_{-1}) \longrightarrow \cdots$$

commute. □

The limited naturality of an H_B^* - display is perhaps less than satisfactory, but it will suffice for our present applications. More "naturality" can be obtained, but apparantly only at the expense of imposing more stringent restrictions on the spaces in the display, see for example [43].

Finally, we shall need to know that if we start with a pointed fibration over B, (f,s), that it is possible to obtain an H_B^*-display of (f,s)

$$(f,s) \longrightarrow (h_0,u_0) \longrightarrow (f_{-1}, s_{-1}) \longrightarrow \cdots$$

wherein all of the pointed spaces

$$(f_{-i}, s_{-i}) \quad : i = 1,2,\ldots \ldots$$

$$(h_{-i}, u_{-i}) \quad : i = 0, 1, \ldots$$

are fibrations over B. This is immediately checked for the explicit construction of a display given in 4.3. We record this additional fact for later use.

<u>Proposition 4.8</u> : Let B, k, H_B^* be as above. Suppose that we are given a pointed fibration over B, (f,s). Then there exists an H_B^* - display

$$(f,s) \longrightarrow (h_0, u_0) \longrightarrow (f_{-1}, s_{-1}) \longrightarrow \cdots$$

where the pointed spaces over B, (f_{-i}, s_{-i}), (h_{-i}, u_{-i}), $i = 0, 1, \ldots$ are all fibrations over B. □

§ 5. The Spectral Kunneth Theorem on $(Top/B)_*$ for Classical Cohomology Theories. The Construction.

Our objective in this section is to show how the elementary machinary of the previous section may be applied to obtain the desired Kunneth theorem. We will begin with the most elementary case.

Proposition 5.1 : Let B, k and $H^*_B\cdot$ be as above. Moreover assume that B is 1-connected. Suppose that $(f,s) \varepsilon \operatorname{obj}(Fib/B)_*$ and that $H^*_B(f,s)$ is projective of finite type. Then for any $(g,t) \varepsilon \operatorname{obj}(Top/B)_*$ with $H^*_B(g,t)$ of finite type the exterior product

$$H^*_B(f,s) \otimes H^*_B(g,t) \longrightarrow H^*(f \wedge_B g,\ s \wedge_B t)$$

induces an isomorphism

$$H^*_B(f,s) \otimes_{H^*(S^o_B,\ s^o_B)} H^*(g,t) \longrightarrow H^*(\ f \wedge_B g,\ s \wedge_B t).$$

Proof: The argument is similar to [27; 6.1-6.3]. First we consider the fibre square

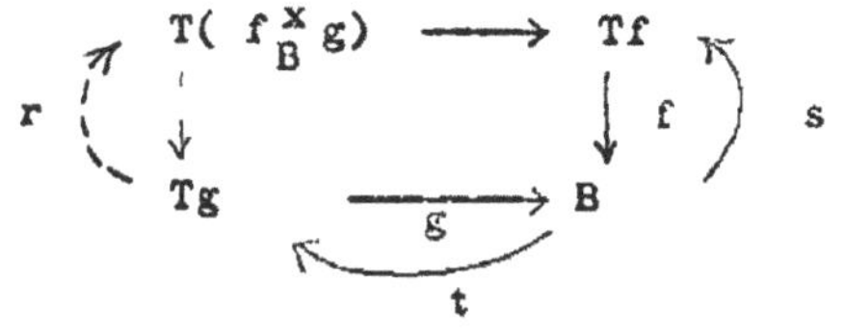

We note that the cross-section

$$s : B \longrightarrow Tf$$

induces a section

$$r : Tg \longrightarrow T(f \times_B g)$$

and that the composite

$$r \cdot t \ ; B \longrightarrow T(f \times_B g)$$

is $s \times_B t$.

Now consider the Serre spectral sequence for the fibration pair

$$\begin{array}{c}(Tf\ ,\ sB)\\ \Big\downarrow f\\ B\ .\end{array}$$

It has

$$E_r \Longrightarrow H^*(Tf\ ,\ sB)$$

$$E_2 = H^*(F_bf,sb) \underset{k}{\otimes} H^*(B)$$

Using the fact that $H^*(Tf,sb)$ is a free $H^*(B)$ - module it may be shown (see appendix to this section) that $E_2 = E_\infty$. Hence the fibration pair (Tf,sB) is totally non-homologus to zero.

From the commutative diagram

$$\begin{array}{ccc} H^*(F_bf,sb) & \longleftrightarrow & H^*(F_bf,sb) \\ \uparrow & & \uparrow \text{ onto} \\ H^*(T(f \underset{B}{\times} g)\ ,\ rTg) & \longleftarrow & H^*(Tf,sB) \end{array}$$

we learn that the fibration pair

$$\begin{array}{c}(T(f \underset{B}{\times} g),\ rTg)\\ \Big\downarrow\\ Tg\end{array}$$

is totally non-homologus to zero. Thus the Serre spactral sequence $\{\bar{E}_r\ ,\ \bar{d}_r\}$ of this fibration pair also collapses. Thus

$$E_0H^*(T(f \underset{B}{\times} g),rTg) = H^*(Tg) \otimes H^*(F_bf,sb)$$

Therefore since

$$E_0\{H^*(Tg) \otimes_{H^*(B)} H^*(Tf,sB)\} = H^*(Tg) \otimes H^*(F_bf,sb)$$

we find by a simple filtration argument that the natural map

$$H^*(Tg) \otimes_{H^*(B)} H^*(Tf,sB) \longrightarrow H^*(T(f \underset{B}{\times} g),rTg)$$

is an isomorphism. From this we deduce that

$$H^*(Tg,tB) \otimes_{H^*(B)} H^*(Tf,sB) \longrightarrow H^*(T(f \underset{B}{\times} g, rTg \cup uTf)$$

is an isomorphism, where $u : Tf \longrightarrow T(f \underset{B}{\times} g)$ is induced by t. One now readily checks that

$$T(f \underset{B}{\times} g)/rTg \cup uTf = T(f \underset{B}{\wedge} g)/s \underset{B}{\wedge} t(B)$$

and the result follows. □

<u>Remark</u>: Note that the only place that we used the one-connectivity of B was in the splitting of the E_2 terms of the Serre spectral sequence into a tensor product. Clearly all we really need to assume is that the local coefficent system in $f : Tf \longrightarrow B$ is trivial.

The Kunneth spectral sequence for classical cohomology theories on $(Top/B)_*$ may now be constructed from the exact couple of a suitable filtered pointed space over B. More precisely, let $(f,s) \in obj(Fib/B)_*$ and $(g,t') \in obj(Top/B)_*$. Choose an H^*_B - display

$$(f,s) \longrightarrow (h_0,u_0) \longrightarrow (f_{-1},s_{-1}) \longrightarrow \cdots$$

by fibrations over B. This is possible by 4.8. For each integer $n \geq 0$ we may then form the filtered pointed space over B

$$(g,t) \hat{\wedge}_B (f_{-n}, s_{-n}) \hookrightarrow \cdots \hookrightarrow (g,t) \hat{\wedge}_B (S_B^n f, S_B^n s) .$$

The H_B^* - exact couple of this filtered object will give us a portion of the desired spectral sequence after a minor reindexing. By increasing n we obtain increasing portions of the spectral sequence. The method is analagous to the construction of the Adams spectral sequence [25] . We proceed to the details.

<u>Proposition 5.2</u>: Suppose that (g,t) and (h,u) are pointed spaces over B.. Then there is a homotopy equivalence

$$(h,u) \hat{\wedge}_B (S_B g, S_B t) \sim S_B(h \hat{\wedge}_B g , u \hat{\wedge}_B t)$$

over B.

<u>Proof</u>: This is merely a matter of writting down the definitions and observing that the standard equivalence for pointed spaces [50]

$$X \wedge S^1 \wedge Y \sim S^1 \wedge (X \wedge Y) \qquad \text{(shuffle)}$$

works equally well for spaces over B . □

<u>Construction</u>: Suppose that $(f,s) \in obj(Fib/B)_*$ and $(g,t) \in obj(Top/B)_*$. Form an H_B^* - display of (f,s) by fibrations over B , say

$$(f,s) \longrightarrow (h_0,u_0) \longrightarrow (f_{-1},s_{-1}) \longrightarrow \cdots .$$

For each integer n form the filtered space

$$(g \hat{\wedge}_B f_{-n}, t \hat{\wedge}_B s_{-n}) \quad > \cdots \quad > (g \hat{\wedge}_B S_B^n f , t \hat{\wedge}_B S_B^n s) .$$

Introduce an exact couple $\mathcal{C}(n)$ by setting

$$D^{-p,q} = H_B^{-p+q-1}(g \hat{\wedge}_B S_B^{n-p} f_{-p} , t \hat{\wedge}_B S_B^{n-p} s_{-p})$$

$$E^{-p,q} = H_B^{-p+q}((g,t) \hat{\wedge}_B (S_B^{n-p} f_{-p} , S_B^{n-p-1} f_{-p-1} , S_B^{n-p} s))$$

where we have extended the notation for smash products over B to pairs in the obvious way. Note the slight index shift in the definition of $D^{-p,q}$. This is to force the couple to adhere to the usual indexing conventions of a cohomology couple [26]. The maps of the couple arise from the exact triangle of the pairs

$$(g,t) \hat{\wedge}_B (S_B^{n-i} f_{-i}, S_B^{n-i-1} f_{-i-1} , S_B^{n-i} s_{-i})$$

in the usual fashion.

As the pair (f,s) , (g,t) and the chosen H_B^* - display of (f,s) will remain fixed throughout our present discussion we will not bother to note the dependence of the spectral sequence of the above couple on these choices and write simply $\{ E_r(n) , d_r(n) \}$ for the spectral sequence associated to this couple.

The naturality of the suspension isomorphism and 5.2 show that we have a natural isomorphism

$$E_r^{p,q}(n) \cong E_r^{p,q+1}(n+1) \quad ; p > 1-n$$

under which

$$d_r^{p,q}(n) = d_r^{p,q+1}(n+1) \quad ; p > 1-n .$$

Thus we may define

$$E_r^{p,q}((g,t),(f,s)) = E_r^{p,q+n}(n) \quad ; \quad p > 1-n .$$

This is independent of the choice of n. Likewise we set

$$d_r^{p,q}((g,t),(f,s)) = d_r^{p,q+n}(n) \quad ; \quad p > 1-n ,$$

which is again independent of n.

It is straight forward to check that with the above definitions $\{ E_r((g,t),(f,s) , d_r((g,t),(f,s)) \}$ is a second quadrant cohomology spectral sequence. This spectral sequence will presently be seen to be the sought for Kunneth spectral sequence for the pointed spaces over B (f,s), (g,t).

<u>Identification of E_2</u>: We turn now to the identification of the E_2 term of the spectral sequence we constructed above. As we shall wish to apply 5.1 at a crucial point we will therefore assume that B is a 1-connected space and that $H^*_B(f,s)$ and $H^*_B(g,t)$ are of finite type. We note that the explicit construction used in 4.3 then shows that

$$H^*_B(S^{n-i}_B f_{-i}, S^{n-i-1}_B f_{-i-1}, S^{n-i}_B s_{-i}) \quad : \quad i = 0, \ldots, n$$

are also of finite type for $n = 0, 1, \ldots$.

Consider the exact sequence

$$0 \longleftarrow H^*_B(S^n_B f, S^n_B s) \longleftarrow H^*_B(S^n_B f, S^{n-1}_B f_{-1}, S^n_B s) \longleftarrow \cdots$$

$$\cdots \longleftarrow H^*_B(S^{n-i}_B f_{-i}, S^{n-i-1}_B f_{-i-1}, S^{n-i}_B s_{-i}) \longleftarrow \cdots$$

$$\cdots \longleftarrow H^*_B(S_B f_{-n+1}, f_{-n}, S_B s_{-n+1})$$

obtained in Proposition 4.2 . Note that this is a partial projective resolution of $H^*_B(S^n_B f, S^n_B s)$ as an $H^*_B(S^0_B, s^0_B)$ - module of length n . From 5.1 we therefore obtain natural isomorphisms

$$\begin{array}{c} H^*_B(g,t) \otimes_{H^*_B(S^0_B, s^0_B)} H^*(S^{n-i}_B f_{-i}, S^{n-i-1}_B f_{-i-1}, S^{n-i}_B s_{-i}) \\ \big\downarrow \\ H^*((g,t) \wedge_B (S^{n-i}_B f_{n-i}\; S^{n-i-1}_B f_{n-i-1}, S^{n-i}_B s_{-i})) \end{array}$$

given by the exterior product.

From the definition of the spectral sequence $\{E_r(n), d_r(n)\}$

we find that $E_2(n)$ is the homology of the complex

$$0 \longleftarrow H_B^*((g,t) \hat{\wedge}_B (S_B^n f, S_B^{n-1} f_{n-1}, S_B^n s)) \longleftarrow \ldots$$

$$\ldots \longleftarrow H_B^*((g,t) \hat{\wedge}_B (S_B^{n-i} f_{n-i}, S_B^{n-i-1} f_{-i-1}, S_B^{n-i} f_{-i})) \longleftarrow \ldots$$

$$\ldots \longleftarrow H_B^*((g,t) \hat{\wedge}_B (S_B f_{-n+1}, f_{-n}, S_B s_{-n+1})) \longleftarrow 0 \ .$$

From our above discussion we find that this complex is isomorphic to the complex

$$0 \longleftarrow H_B^*(g,t) \otimes_{H_B^*(S_B^0, s_B^0)} H_B^*(S_B^n f, S_B^{n-1} f_{n-1}, S_B^n s)) \longleftarrow \ldots$$

$$\ldots \longleftarrow H_B^*(g,t) \otimes_{H_B^*(S_B^0, s_B^0)} H_B^*(S_B^{n-i} f_{-i}, S_B^{n-i-1} f_{-i-1}, S_B^{n-i} s_{-i}) \longleftarrow \ldots$$

$$\ldots \longleftarrow H_B^*(g,t) \otimes_{H_B^*(S_B^0, s_B^0)} H_B^*(S_B f_{-n+1}, f_{-n}, S_B s_{-n+1}) \longleftarrow 0 \ .$$

From the definition of the functor Tor and our discussion above we thus find

$$E_2^{p,q}(n) = \mathrm{Tor}^{p,q}_{H_B^*(S_B^0, s_B^0)}(H_B^*(g,t) \ , \ H_B^*(S_B^n f, S_B^n s)) \quad ; \ p > 1-n \ .$$

From the suspension isomorphism

$$H_B^*(S_B^n f, S_B^n s) = s^n H_B^*(f,s)$$

we thus obtain for $p > 1-n$ the isomorphism

$$E_2^{p,q+n}(n) = \mathrm{Tor}^{p,q}_{H_B^*(S_B^0, s_B^0)}(H_B^*(g,t), \ H_B^*(f,s)) \ .$$

Thus taking into account the dimension shifts in the definition of $\{ E_r((g,t) \ ,(f,s)) \ , \ d_r((g,t), \ (f,s))\}$ we find that

$$E_2^{p,q}((g,t),(f,s) = \mathrm{Tor}^{p,q}_{H_B^*(S_B^0, s_B^0)}(H_B^*(g,t), H_B^*(f,s))$$

for all (p,q) .

<u>Naturality</u>: So far we have shown how, given a particular

H_B^* - display of (f,s) we may construct a spectral sequence with some of the required properties to be a Kunneth spectral sequence in terms of our previous discussion. We must show that our construction is independent of the particular choice of display made and is functorial. We thus suppose given $(g't')$, $(g",t") \in obj(Top/B)_*$ and (f',s'), $(f",s") \in obj(Fib/B)_*$, together with morphisms

$$\varphi : (g',t') \longrightarrow (g",t") \quad , \quad \psi : (f',s') \longrightarrow (f",s")$$

in $(Top/B)_*$. We will assume that all cohomologies that occur in the sequel are of finite type. (The explicit constructions 4.2 and 4.5 are needed at this point.)

We also assume given H_B^* - displays by fibrations

$$(f',s') \longrightarrow (h_0',u_0') \longrightarrow (f_{-1}',u_{-1}') \longrightarrow \cdots$$

$$(f",s") \longrightarrow (h_0",u_0") \longrightarrow (f_{-1}",u_{-1}") \longrightarrow \cdots$$

from which we have constructed the spectral sequences $\{E_r((g',t'),(f',s')), d_r((g',t'),(f',s'))\}$ and $\{E_r((g",t"),(f",s")), d_r((g",t"),(f",s"))\}$ by the method described above.

According to 4.7 we may construct a diagram of fibrations over B

$$\begin{array}{ccccccccc}
(f',s') & \longrightarrow & (h_0',u_0') & \longrightarrow & (f_{-1}',s_{-1}) & \longrightarrow & (h_{-1}',u_{-1}') & \longrightarrow & \cdots \\
\| & & \uparrow & & \uparrow & & \uparrow & & \\
(f',s') & \longrightarrow & (\bar{h}_0,\bar{u}_0) & \longrightarrow & (\bar{f}_{-1}',\bar{s}_{-1}') & \longrightarrow & (\bar{h}_{-1}',\bar{u}_{-1}') & \longrightarrow & \cdots \\
\psi\downarrow & & \downarrow & & \downarrow & & \downarrow & & \\
(f",s") & \longrightarrow & (h_0",u_0") & \longrightarrow & (f_{-1}",s_{-1}") & \longrightarrow & (h_{-1}",u_{-1}") & \longrightarrow & \cdots
\end{array}$$

where the middle row is also an H_B^* - display.

From this diagram we may construct for each non-negative

integer n a commutative diagram

$$\begin{array}{ccccc}
(g' \wedge_B f'_{-n}, t' \wedge_B s'_{-n}) & \longrightarrow & \cdots & \longrightarrow & (g' \wedge_B S^n_B f', \; t' \wedge_B S^n_B \; s') \\
\uparrow & & & & \| \\
(g' \wedge_B \bar{f}'_{-n}, t' \;{}_B\; \bar{s}'_{-n}) & \longrightarrow & \cdots & \longrightarrow & (g' \wedge_B S^n_B f', \; t' \wedge_B S^n_B \; s') \\
\downarrow & & & & \downarrow \\
(g'' \wedge_B f''_{-n}, t'' \wedge_B s''_{-n}) & \longrightarrow & \cdots & \longrightarrow & (g'' \wedge_B S^n_B f'', \; t'' \wedge_B S^n_B \; s'')
\end{array}$$

of filtered pointed spaces over B. The top two rows provide us with spectral sequences that we denote by $\{E'_r(n) \; , \; d'_r(n)\}$ and $\{\bar{E}'_r(n) \; , \; \bar{d}'_r(n)\}$. The top maps of pointed filtered spaces over B provide morphisms of spectral sequences

$$\{\bar{E}'_r(n), \bar{d}'_r(n)\} \longrightarrow \{E'_r(n), d'_r(n)\} \qquad : n \geq 0 \; .$$

Upon the E_2 term we find from our previous discussion the above map induces an isomorphism

$$\begin{array}{ll}
\bar{E}'^{p,q}_2 = \mathrm{Tor}^{p,q}_{H^*_B(S^0_B, s^0_B)}(H^*_B(g',t'), H^*_B(f',s')) & \\
\downarrow \cong & \qquad p > 1-n. \\
E'^{p,q}_2 = \mathrm{Tor}^{p,q}_{H^*_B(S^0_B, s^0_B)}(H^*_B(g',t'), H^*_B(f',s')) &
\end{array}$$

Thus appealing to the definitions we find that the spectral sequences $\{ E_r((g',t'),(f',s')), \; d_r((g',t'),(f',s'))\}$ and $\{ \bar{E}_r((g',t'),(f',s'), \; d_r((g',t'),(f',s'))\}$ are naturally isomorphic from the term E_2 on.

With this in mind the lower map of pointed filtered spaces over B provides a morphism of spectral sequences

$$\{ E''_r(n) \; , \; d''_r(n)\} \longrightarrow \{\bar{E}'_r(n), \bar{d}'_r(n)\}$$

which on the term E_2 yields the commuattive diagram

$$\begin{array}{ll}
E''^{p,q+n}_2(n) = \mathrm{Tor}^{p,q}_{H^*_B(S^0_B, s^0_B)}(H^*_B(g'',t''), H^*_B(f'',s'')) & \\
p > 1-n \; \downarrow \qquad\qquad\qquad \downarrow & \mathrm{Tor}(H^*_B(\varphi) \; , \; H^*_B(\psi)) \\
\bar{E}^{p,q+n}_2(n) = \mathrm{Tor}^{p,q}_{H^*_B(S^0_B, s^0_B)}(H^*_B(g',t'), H^*_B(f',s')) &
\end{array}$$

Thus from stability and our definitions we may assemble these maps into a morphism of spectral sequences

$$\{E_r((g",t"),(f",s")\ ,\ d_r((g",t"),(f",s"))\}$$

$$\Big\downarrow \{E_r(\varphi\ ,\psi\ )\}$$

$$\{\ E_r((g',t'),(f',s'))\ ,\ d_r((g',s'),(f',s')).\}$$

On the term E_2 we find

$$E_2(\varphi\ ,\psi\ )\ =\ \mathrm{Tor}_{H_B^*(S_B^0,s_B^0)}(H_B^*(\varphi\ )\ ,\ H_B^*(\psi\ ))$$

which establishes the naturality of our construction

<u>Convergence</u>: In order to establish the convergence of the spectral sequence we will appeal to the criterion of [11;XV.4]. To this end let us suppose that

$$(f,s) \longrightarrow (h_0,u_0) \longrightarrow (f_{-1},s_{-1}) \longrightarrow (h_{-1}u_{-1}) \longrightarrow \dots$$

is a dispaly <u>constructed according to the algorithm given in 4.3</u> . We claim that

$$H_B^j(f_{-p},s_{-p})\ =\ 0 \quad \text{for} \quad j < 2p\ .$$

To see this let us examine in detail the cohomology morphism induced by the natural map

$$(f,s) \longrightarrow \Gamma\ \phi(f,s)$$

which figures so prominently in the proof of 4.3 and 4.4.

We note that

$$T(\Gamma\ \phi)(f,s)\ =\ Tf/sB \times B$$

the morphism

$$s:\ B \longrightarrow T\Gamma\phi(f,s)$$

being given by

$$x \longrightarrow (sB/sB\ ,\ x\ )\ ,$$

Thus we find that

$$H_B^*(\Gamma\phi(f,s)) = H^*(Tf/sB \times B\ ,\ * \times B)$$
$$= \tilde{H}^*(Tf/sB) \otimes H^*(B)$$

while

$$H_B^*(f,s) = \tilde{H}^*(Tf,sB)$$

and the induced map

$$H_B^*(\Gamma\phi(f,s)) \longrightarrow H_B^*(f,s)$$

coincides with the cup-product mapping

$$H^*(B) \otimes \tilde{H}^*(Tf/sB) \longrightarrow \tilde{H}^*(Tf/sB)\ .$$

We are now ready to bring into play our assumption that B is one-connected. For since this is the case it follows that $\tilde{H}^0(B) = 0 = \tilde{H}^1(B)$. Therefore, if

$$\tilde{H}^j(Tf/sB) = 0 \quad \text{for} \quad j < c$$

then

$$[H^*(B) \otimes \tilde{H}^*(Tf/sB)]^j \longrightarrow \tilde{H}^j(Tf/sB)$$

is seen to be an isomorphism for $j \leq c + 1$. From the exact sequence of the pair

$$\varphi:\ (f,s) \longrightarrow \Gamma\phi(f,s)$$

it readily follows that

$$H_B^j(TC_B(\varphi)\ ,\ s_{C_B(\varphi)}) = 0$$

for $j > c + 2$.

Applying these observations inductively, starting with $c = 0$, yields our assertion

$$H_B^j(f_{-p},s_{-p}) = 0 \quad \text{for} \quad j < 2p.$$

To see how this applies to our convergence problem let us examine the filtration

$$F_{-p}H_B^j(g \wedge_B f, t \wedge_B s) = \mathrm{Im}\left\{\begin{array}{c} H_B^{j+n}(g \wedge_B S_B^{n-p} f_{-p}, t \wedge_B S_B^{n-p} s) \\ \downarrow \\ H_B^{j+n}(g \wedge_B S_B^n f_{-p}, t \wedge_B S_B^n s) \end{array}\right.$$

Now as $H_B^j(f_{-p}, s_{-p}) = 0$ for $j < 2p$ one finds by stability that

$$H_B^{j+n}(S_B^{n-p} f_{-p}, S_B^{n-p} s_{-p}) = 0$$

for $j < p$. Hence it may be shown (see appendix to this section) that

$$H_B^{j+n}(g \wedge_B S_B^{n-p} f_{-p}, t \wedge_B S_B^{n-p} s) = 0$$

for $j < p$. Hence of necessity

$$F_{-p}H_B^j(g \wedge_B f, t \wedge_B s) = 0$$

for $p > j$. Thus the fåltration is finite in each degree and by [11;XV.4] it follows that we have shown that the spectral sequence $\{E_r((g,t),(f,s), d_r((g,t),(f,s))\}$ converges strongly to $H_B^*(g \wedge_B f, s \wedge_B t)$. Hence we have established:

<u>Theorem 5.3</u>: Let B be a 1- connected space and k a field, with $H_B^*(\ )$ denoting the prolongation of $H^*(\ ;k)$ to a cohomology theory on (Top/B)$_*$.

Suppose that $(f,s) \in \mathrm{obj}(\mathrm{Fib}/B)_*$ and $(g,t) \in \mathrm{obj}(\mathrm{Top}/B)_*$ with $H_B^*(f,s)$ and $H_B^*(g,t)$ of finite type.

Then there exists a natural spectral sequence of cohomology type and concentrated in the second quadrant, $\{E_r((g,t),(f,s)$, $d_r((g,t),(f,s))\}$, such that

(1) $E_r((g,t),(f,s)) \Longrightarrow H_B^*(g \wedge_B f, t \wedge_B s)$

the convergence being in the naive sense;

(2) $E_2^{p,q}((g,t),(f,s) = \mathrm{Tor}_{H_B^*(S_B^0, s_B^0)}^{p,q}(H_B^*(g,t), H_B^*(f,s))$;

(3) the edge homomorphism

$$H_B^*(g,t) \otimes_{H_B^*(S_B^0,s_B^0)} H_B^*(f,s) = E_2^{0,*}((g,t),(f,s)) \longrightarrow E_\infty^{0,*}((g,t),(f,s)) \hookrightarrow H_B^*(g \hat{\wedge}_B f, t \hat{\wedge}_B s)$$

coincides with the exterior product.

$$H_B^*(g,t) \otimes_{H_B^*(S_B^0,s_B^0)} H_B^*(f,s) \longrightarrow H_B^*(g \hat{\wedge}_B f, t \hat{\wedge}_B s) ;$$

Suppose (g',t'), $(g'',t'') \in \mathrm{obj}(\mathrm{Top}/B)_*$, (f',s'), (f'',s'') $\in \mathrm{obj}(\mathrm{Fib}/B)_*$ $\varphi: (g',t') \longrightarrow (g'',t'')$ and $\psi:(f',s') \longrightarrow (f'',s'')$ are morphisms in $(\mathrm{Top}/B)_*$ and $H_B^*(g',t')$, $H_B^*(g'',t'')$, $H_B^*(f',s')$ and $H_B^*(f'',s'')$ are of finite type, then

(4) the induced map of spectral sequences has

$$E_2(\varphi, \psi) = \mathrm{Tor}_{H_B^*(S_B^0,s_B^0)}(H_B^*(g,t),H_B^*(f,s))$$

under the identifications of (2).

<u>Proof</u>: All that remains to be established is (3) which is elementary and left to the reader. □

As we have used the multiplicative structure of the cohomology theory H_B^* in setting up the Kunneth spectral sequence it is natural to expect this multiplicative structure to be reflected in the spectral sequence. This is indeed the case and straightforward to establish but long and tedious. The interested reader is referred to [43] for the details.

We turn now to the proofs of two technical results that we employed above. The proof are quite easy and are appended to this section.

Appendix to Section 5: Some Technical Lemmas.

Our intention in this section is to discuss some of the technical points from the precceding section whose proofs were deferred so as to not interrupt the flow of the arguments. The first of these points is embodied in :

Proposition A.5.1: Suppose that B is a 1- connected space and $\pi: E \longrightarrow B$ is a fibration. Let $E_0 \subset E$ be a subspace with $\pi|E_0 : E_0 \longrightarrow B$ also a fibration. Denote by $\{E_r, d_r\}$ the $H^*(\ ;k)$ - cohomology spectral sequence of the fibration pair $\pi: (E, E_0) \longrightarrow B$. Then the following two conditions are equivalent:

(1) $H^*(E,E_0;k)$ is a free $H^*(B;k)$ - module,

(2) $E_2 = E_\infty$.

Proof: Let us first consider the implication (2) $\Longrightarrow$ (1). Let F be the fibre of π and F_0 the fibre of $\pi|E_0$. Then

$$E_2 = H^*(F,F_0) \otimes H^*(B)$$

(where we have abreviated $H^*(\ ;k)$ to $H^*(\ )$,) and hence is a free $H^*(B)$ - module. As $E_2 = E_\infty$ it follows that there is a filtration of $H^*(E,E_0)$ by $H^*(B)$ - submodules whose associated graded is a free $H^*(B)$ - module. An easy induction over the filtration now establishes that $H^*(E,E_0)$ is a free $H^*(B)$ - module. (Compare for example with [38;5.9].) Thus (2) $\Longrightarrow$ (1).

Let us consider now the reverse implication. We assume that $H^*(E,E_0)$ is a free $H^*(B)$ - module. If $E_0H^*(E,E_0)$ denotes the associated graded under the Serre filtration, then this too is seen to be a free $H^*(B)$ - module. Now suppose to the contrary that $E_2 \neq E_\infty$. Let $u \in H^*(F,F_0)$ be a class of

lowest dimension not in the image of $H^*(E,E_0) \longrightarrow H^*F,F_0)$.
Then there is an integer r such that

$$d_r(u) = \sum v_i \otimes b_i$$

with

$$v_i \in H^*(F,F_0) \quad ; \deg v_i < \deg u$$
$$b_i \in H^*(B) \ .$$

It readily follows that v_i and b_i survive to E_∞ for dimension reasons(recall how we chose u) , to yield the non-trivial relation

$$\sum v_i \otimes b_i = 0 \quad \varepsilon \ E_\infty$$

contrary to the freeness of E_∞ , and hence we must have $E_2 = E_\infty$ as required. □

We turn now to the second technical result that we need. This is contained in:

<u>Proposition A.5.2</u>: Suppose that $(f,s) \in obj(Fib/B)_*$ and $(g,t) \in obj(Top/B)_*$. Moreover suppose that

$$H^j_B(f,s) = 0$$

for $j < c$. Then

$$H^j_B(g \wedge_B f, \ t \ _B \ s) = 0$$

for $j < c$.

<u>Proof</u>: Consider the fibre square

$$\begin{array}{ccc} T(f \underset{B}{\times} g) & \longrightarrow & Tf \\ h\downarrow & & \downarrow f \\ Tg & \underset{g}{\longrightarrow} & B \end{array}$$

From the condition that $H^j_B(f,s) = 0$ for $j < c$ and the exact cohomology sequence of the pair (Tf,sB) we find that

$$s^* : H^*(Tf) \longrightarrow H^*(sB)$$

is an isomorphism for $* < c$. # This in turn implies that

$$f^* : H^*(B) \longrightarrow H^*(Tf)$$

is an isomorphism for $* < c$. Thus no element of $H^*(F(f))$ of degree $< j-1$ has a non zero coboundary in the Serre spectral sequence of f. Thus the same is true for the induced fibration h, i.e. no element of $H^*(F(f)) = H^*(F(h))$ can have a non-zero coboundary in the Serre spectral sequence of the fibration h. This of course means that

$$h^* : H^*(Tg) \longrightarrow H^*(T(g \underset{B}{\times} f)$$

is an isomorphism for $* < j$. Thus the natural map

$$\eta : f \underset{B}{\curlyvee} g \longrightarrow f \underset{B}{\times} g$$

induces an isomorphism of H_B^* - cohomology for $* < c$. As the cofibre (in the category $(Top/B)_*$!) of η is $f \wedge_B g$ the result follows. □

6. The Eilenberg-Moore Spectral Sequence.

In this section we shift the emphasis away from the Kunneth spectral sequence on $(Top/B)_*$ to the Eilenberg-Moore spectral sequence for fibre squares. We will first show how the Eilenberg-Moore spectral sequence may be regarded as a special case of the spectral Kunneth theorem for the classical cohomology theoreis on $(Top/B)_*$. We will then describe various functorial and multiplicative properties of the spectral sequence that we will

---

Do not forget that s is a cross section and hence s^* is always epic.

use in the applications of the next section, but whose proof we leave to the reader. (See also [43].) We will close the section by indicating several elementary extensionsof the theory.

We continue to adhere to the notations and conventions established in the previous section. Thus k continues to denote a fixed field and we abreviate $H^*(\ ; k)$ to $H^*(\ )$, etc.

Theorem 6.1: Let k be a field and B a 1-connected space. Suppose that

$$f' : Tf' \longrightarrow B \longleftarrow Tf'' : f''$$

are spaces over B and that f" is a fibration. Assume that $H^*(B)$, $H^*(Tf')$ and $H^*(Tf'')$ are of finite type.

Under these conditions there exists a second quadrant cohomology spectral sequence $\{E_r(f',f''),d_r(f',f'')\}$ with the following properties:

(1) $E_r(f',f'') \Longrightarrow H^*(Tf' \times_B Tf'')$, the convergence being in the naive sense;

(2) $E_2(f',f'') = \mathrm{Tor}_{H^*(B)}(H^*(Tf'), H^*(Tf''))$

(3) the edge homomorphism

$$\begin{array}{ccc} H^*(Tf') \otimes_{H^*(B)} H^*(Tf'') = E_2^{0,*}(f',f'') & & \\ \downarrow \text{edge} & & \\ E_\infty^{0,*}(f',f'') \hookrightarrow H^*(Tf' \times_B Tf'') & & \end{array}$$

coincides with the natural map induced by the exterior cross-product;

(4) if g' , g" are spaces over B satisfying the above conditions and

$$\varphi' : g' \longrightarrow f' \ , \quad \varphi'' ; g'' \longrightarrow f''$$

are morphisms of spaces over B, then under the identification of (2) we have

$$E_2(\varphi', \varphi'') = \mathrm{Tor}_{H^*(B)}(\varphi^{*\prime}, \varphi^{*\prime\prime})$$

where $E_r(\varphi', \varphi'')$ is the morphism of spectral sequences induced by (φ', φ'').

<u>Proof</u>: As in section 2 let

$$G : \mathrm{Top}/B \longrightarrow (\mathrm{Top}/B)_*$$

be the adjoint of the forgetful functor. Note that

$$G(f' \underset{B}{\times} f'') = G(f') \underset{B}{\wedge} G(f'') .$$

We also have that

$$H_B^*(G(f)) = H^*(T(f)) \quad , \quad f \in \mathrm{obj}\ \mathrm{Top}/B \quad .$$

In particular

$$H_B^*(S_B^0, s_B^0) = H^*(B) .$$

If we now make these substitutions into the Kunneth spectral sequence (5.3) the result follows. □

In the sequel we shall need to make use of the relation between the spectral sequence of 6.1 (which we shall refer to as the Eilenberg-Moore spectral sequence) and the multiplicative properties of the cohomology theory $H^*(\ )$. Due to lack of time in the original lectures the details of the proofs of these additional properties were not presented. As in the original lectures they are left s exercises for the reader.

<u>Continuation of 6.1</u>: The spectral sequence $\{E_r(f', f''), d_r(f', f'')\}$ is moreover a spectral sequence of algebras, that is,

(5) for each integer $r \geq 2$ (E_r, d_r) is a bigraded differential algebra and

$$E_{r+1} = H(E_r, d_r)$$

as bigraded algebras;

(6) the convergence in (2) is as algebras, i.e. the filtration $\{ F^{-p}H^*(Tf' \underset{B}{\times} Tf'')\}$ is compatable with the algebra structure and

$$E^0H^*(Tf' \underset{B}{\times} Tf'') = E_\infty$$

as bigraded algebras. $\square$

The interested reader may wish to consult [43] where the connection between the spectral sequence and the Steenrod algebra (when $k = \mathbb{Z}_p$; p a prime) is developed. While not exploited in the present lectures this result is of interest in connection with the study of the differentials of the spectral sequence (see for example [38]) . The study of the differentials is carried further in [46] on the basis of the results established in [43].

In our discussion of naturality of the spectral sequence we did not discuss the functorial dependence of the spectral sequence on B . This is essential to complete our analogy with the work of Eilenberg and Moore and allows us to introduce the fibre square approach which has proven so important for the applications. The details of the requesite naturality arguments are straight forward and as in the original lectures are left to the reader.

<u>Definition</u>: A commuative square of topological spaces

$$\begin{array}{ccc} E & \longrightarrow & Y \\ h \downarrow & & \downarrow g \\ X & \xrightarrow{f} & B \end{array} \qquad \mathcal{F}$$

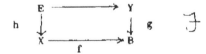

is called a fibre square iff

(1) $g : Y \longrightarrow B$ is a fibration and

(2) $h : E \longrightarrow X$ is the fibration induced from g by the map f.

If in addition

(3) B is 1-connected

then we say that $\mathfrak{F}$ is 1-connected. If in addition

(4) $H^*(X)$, $H^*(B^*$, $H^*(Y)$ are of finite type

then we say that $\mathfrak{F}$ is of finite type (with respect to k).

If $\mathfrak{F}'$ and $\mathfrak{F}''$ are fibre squares then a morphism $\varphi : \mathfrak{F}' \longrightarrow \mathfrak{F}''$ of fibre squares is a commutative cube

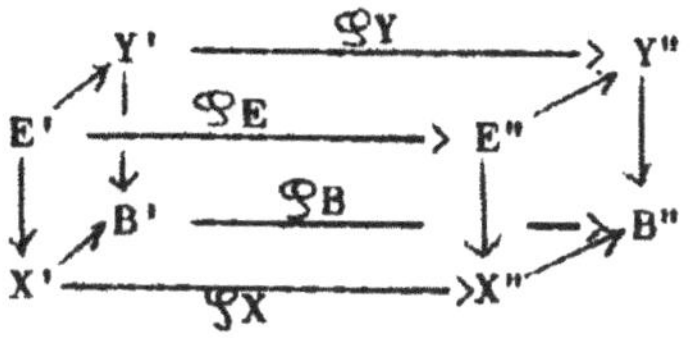

Evidently compositions will behave well and we have the category of fibre squares, which we denote by $\mathfrak{F}Sq$.

<u>Theorem 6.2</u>: Let be a 1-connected fibre square

of finite type with respect to the field k. Then there exists a second quadrant cohomology spectral sequence $\{E_r(\mathfrak{F}), d_r(\mathfrak{F})\}$ with the following properties:

(1) $E_r(\mathfrak{F}) \Longrightarrow H^*(E)$ the convergence being in the naive sense;

(2) $E_2^{s,t}(\mathfrak{F}) = \mathrm{Tor}^{s,t}_{H^*(B)}(H^*(X), H^*(Y))$;

(3) the edge homomorphism

$$H^*(X) \otimes_{H^*(B)} H^*(Y) = E_2^{0,*}(\mathfrak{F}) \longrightarrow E_\infty^{0,*}(\mathfrak{F}) \hookrightarrow H^*(E)$$

coincides with the composition

$$H^*(X) \otimes_{H^*(B)} H^*(Y) \longrightarrow H^*(E) \otimes H^*(E) \xrightarrow{m^*} H^*(E)$$

where m^* is the multiplication map.

Moreover the spectral sequence is a spectral sequence of bigraded commutative algebras, that is,

(4) for each integer $r \geq 2$, (E_r, d_r) is a bigraded commutative differential algebra and

$$E_{r+1} = H(E_r, d_r)$$

as bigraded algebras;

(5) the convergence in (2) is as algebras, i.e. the filtration $\{F^{-p}H^*(E)\}$ is compatable with the algebra structure and

$$E_0H^*(E) = E_\infty(\mathfrak{F})$$

as bigraded algebras.

The spectral sequence, together with the superstructure listed above, is natural with respect to maps of 1-connected fibre squares of finite type. Moreover if $\varphi : \mathfrak{F}' \longrightarrow \mathfrak{F}''$ is a morphism of 1-connected fibre squares of finite type then the induced map of spectral sequences $\{E_r(\varphi)\}$ satisfies

(6) under the identification (2) we have

$$E_2(\varphi) = \mathrm{Tor}_{\varphi_B^*}(\varphi_X^*, \varphi_Y^*) \quad . \quad \square$$

<u>Remarks</u>: (1) While we have not demonstrated that the spectral sequence of 6.1 - 6.2 coincides with that introduced by Eilenberg and Moore in [18] we will still refer to 6.1

-6.2 as the Eilenberg-Moore spectral sequence. As we will never use anything but the fundamental properties 6.1-6.2/ (1) - (6) there will be no real chance for confusion. Presumebly the two spectral sequences coincide altgough this would seem to require some sort of super acyclic models argument or a detailed description of an isomorphism.

(2) The hypothesis that the fibre square be 1- connected is stronger than is needed. It would suffice to assume that $H^*(B)$ acts trivially on the cohomology of the fibre. This assumption suffices for all but the discussion of convergence given above. A more delicate argument coupled with the results of [17] is needed. The details are routine.

CHAPTER II

Multiplicative Fibrations

In this chapter we will present some applications of the Eilenberg-Moore spectral sequence. Perhaps more precisely we should say that we are going to consider some problems where the Eilenberg-Moore spectral sequence may be applied to yield results of particular significance.

Let us begin by describing the problem around which this chapter revolves. (We will assume that the reader is familiar with the terminology and rudimentary properties of Hopf algebras. The classical reference is [30].)

Let π be an abelian group. For each positive integer n there is then a topological abelian monoid $K(\pi, n)$; the Eilenberg-MacLane space of type (π, n), [10],[47],[36]. This space classifies the functor $H^n(- ;\pi)$. The cohomology of such spaces are very important in the study of cohomology operations [10] and have been computed by Cartan [10] and Serre [36]. Being a topological monoid, $H_*(K(\pi, n); k)$ is a Hopf algebra for any field k. Among the results of Cartan and Serre may be found the following general structure theorem for these Hopf algebras.

<u>Theorem</u>(Cartan-Serre): Let π be an abelian group, n a positive integer and p a prime number. Then $H_*(K(\pi, n); \mathbb{Z}_p)$ is a coprimitive bicommutative Hopf algebra.

The problem that we will be concerned with is how does this result generalize to more "complicated" topological monoids. It is clear that the word "complicated" may be interpreted in the sense of homotopy theory. Complication from a homotopy point of view is adequately measured by the number of non-zero homotopy groups, or more precisely by the height of the Postnikov tower [47].

Thus we find that we will have to study homology relations in the Postnikov tower of a topological monoid. This we will do with the aid of the Eilenberg-Moore spectral sequence.

We also find that we will have to generalize the notion of coprimitive bicommutative Hopf algebra. There are two ways in which we might attempt this. The first involves studying how "non-commutative" a given Hopf algebra is and the second involves studying the order of the Frobenius morphism. By analogy with the category of all groups we will introduce a notion of commutator Hopf algebras. This approach may be combined with the Frobenius to yield a p-derived series that is the appropriate generalization that we seek.

We will begin by abstracting a suitable homotopy formulation of the Postnikov tower of a topological monoid and studying homology relations in such a situation. This will entail several digressions on Hopf algebras and computational problems. Along the way to our generalization of the theorem of Cartan and Serre we will encounter several open problems and we will discuss these at the end of the chapter.

Most of the material presented in this chapter is an outgrowth and/or part of my collaboration with J.C.Moore. Many of these results have appeared in [31] although not quite from the present point of view.

§ 1. Hopf Fibre Squares

Let us begin with a definition.

Definition: A Hopf space is a pair (X, μ) where X is a pointed space and $\mu : X \times X \longrightarrow X$ is a map of pointed spaces such that

(1) $\mu(*, x) = x = \mu(x, *)$ for all $x \in X$;

(2) the diagram

$$\begin{array}{ccc} X \times X \times X & \xrightarrow{\mu \times 1} & X \times X \\ {\scriptstyle 1 \times \mu}\downarrow & & \downarrow{\scriptstyle \mu} \\ X \times X & \xrightarrow{\mu} & X \end{array}$$

is homotopy commutative.

Thus in one sense a Hopf space is just a homotopy associative H-space, and in another is just a monoid in the homotopy category $[Top_*]$. Indeed from the point of view of the crude homotopy theory that we will adopt, the Hopf spaces are the homotopy analog of monoids.

If (X', μ') and (X'', μ'') are Hopf spaces, a morphism of Hopf spaces

$$f : (X', \mu') \longrightarrow (X'', \mu'')$$

is a continuous pointed map $f : X' \longrightarrow X''$ such that

$$\begin{array}{ccc} X' \times X' & \xrightarrow{f \times f} & X'' \times X'' \\ \downarrow{\scriptstyle \mu'} & & \downarrow{\scriptstyle \mu''} \\ X' & \xrightarrow{f} & X'' \end{array}$$

is homotopy commutative.

Remarks: (1) Let k be a field and suppose that (X, μ) is a Hopf space. Then

$$\mu_*: H_*(X,k) \otimes_k H_*(X;k) \longrightarrow H_*(X;k)$$

is an associative algebra structure on $H_*(X,k)$. Taken together with the coalgebra structure of $H_*(X;k)$ induced by the diagonal map

$$\Delta : X \longrightarrow X \times X$$

we find that $(H_*(X;k), \mu_*, \Delta_*)$ is a Hopf algebra [30].

If in addition $H_*(X;k)$ is of finite type then the Kunneth theorem works well in cohomology and we find that $(H^*(X;k), \mu^*, \Delta^*)$ is also a Hopf algebra.

These two Hopf algebras are in duality.

Note that the homotopy associative condition on the multiplication $\mu : X \times X \longrightarrow X$ is the most reasonable condition to impose to conclude that $H_*(X;k)$ is a Hopf algebra.

(2) With k still a field let us suppose that

$$f : (X', \mu') \longrightarrow (X"; \mu")$$

is a morphism of Hopf spaces. Then the induced morphism

$$f_*: H_*(X';k) \longrightarrow H_*(X"; k)$$

is a morphism of Hopf algebras. If the appropriate finite type conditions are present then

$$f^*: H^*(X";k) \longrightarrow H^*(X';k)$$

is also a morphism of Hopf algebras. The morphisms f_*, f^* will also be in duality.

(3) Clearly compositions of morphisms of Hopf spaces behave well and we may form the category $\mathcal{H}Sp$ of Hopf spaces.

Notation and Conventions: In the sequel we will not be concerned with the study of a particular multiplication on a Hopf space (X, μ). It will suffice for most of our purposes simply to know that some multiplication has been chosen and/or is preserved by various maps, constructions, etc. We will therefore allow ourselves the liberty of writing X instead of (X, μ) and call X a Hopf-space. This should cause no confusion.

Definition: A Hopf fibre square is a fibre square

$$\begin{array}{ccc} E & \xrightarrow{h} & Y \\ {\scriptstyle \pi}\downarrow & & \downarrow{\scriptstyle g} \\ X & \xrightarrow[f]{} & B \end{array}$$

where

(1) E,X,Y and B are Hopf spaces and

(2) f, g, π, h are morphisms of Hopf spaces.

Morphisms of Hopf fibre squares may be defined in the obvious manner by comparison with I.6. We may then form the category of Hopf fibre squares, which we denote by $\mathcal{HFS}_b$. Much of the sequel will be concerned with the study of homology and cohomology relations in Hopf fibre squares. Our main technical tool will be the Eilenberg-Moore spectral sequence of I.6. We therefore close this section by indicating some additional special features of the Eilenberg-Moore spectral sequence in this case. As is to be expected these take the form of Hopf algebra conditions. These are consequences of naturality and our previous work. We summarize this in:

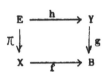

<u>Theorem 1.1</u>: Let $\mathfrak{F}$ be a 1-connected Hopf fibre square,

$$\begin{array}{ccc} E & \longrightarrow & Y \\ \downarrow & & \downarrow g \\ X & \xrightarrow[f]{} & B \end{array}$$

of finite type with respect to $\mathbb{Z}_p$, p a prime. Then there exists a second quadrant cohomology specral sequence $\{E_r(\mathfrak{F}), d_r(\mathfrak{F})\}$ with the following properties:

(1) $E_r(\mathfrak{F}) \Longrightarrow H^*(E;\mathbb{Z}_p)$, the convergence being in the naive sense.

(2) $$E_2^{s,t}(\mathfrak{F}) = \operatorname{Tor}^{s,t}_{H^*(B;\mathbb{Z}_p)}(H^*(X;\mathbb{Z}_p), H^*(Y;\mathbb{Z}_p))$$

(3) The edge homomorphism

$$\begin{array}{lcl} H^*(X;\mathbb{Z}_p) \otimes_{H^*(B;\mathbb{Z}_p)} H^*(Y;\mathbb{Z}_p) = E_2^{0,*}(\mathfrak{F}) & & \\ \qquad\qquad\qquad\qquad\qquad\qquad \downarrow e & & \\ \qquad\qquad\qquad\qquad\qquad\qquad E_\infty^{0,*}(\mathfrak{F}) & \hookrightarrow & H^*(E;\mathbb{Z}_p) \end{array}$$

is the composition

$$\begin{array}{ccc} H^*(X;\mathbb{Z}_p) \otimes_{H^*(B;\mathbb{Z}_p)} H^*(Y;\mathbb{Z}_p) & \longrightarrow & H^*(E;\mathbb{Z}_p) \otimes_{H^*(B;\mathbb{Z}_p)} H^*(E;\mathbb{Z}_p) \\ & & \downarrow \text{multiply} \\ & & H^*(E;\mathbb{Z}_p) \; . \end{array}$$

Moreover, the spectral sequence is a spectral sequence of bigraded commutative Hopf algebras and

(4) the convergence in (1) is as Hopf algebras, i.e., the filtration $\{F^q H^*(E;\mathbb{Z}_p)\}$ is compatible with the Hopf algebra structure and

$$E_0\, H^*(E;\mathbb{Z}_p) \cong E_\infty(\mathfrak{F})$$

as bigraded Hopf algebras;

(5) under the identification (2) the Hopf algebra

structure on $E_2(\mathcal{F})$ coincides with the standard Hopf algebra structure of

$$\operatorname{Tor}_{H^*(B; \mathbb{Z}_p)} (H^*(X; \mathbb{Z}_p), H^*(Y; \mathbb{Z}_p)).$$

Finally for each integer $r \geq 2$, $E_r^{s,}(\mathcal{F})$ is a graded $H^*(B; \mathbb{Z}_p)$-module for all $s \in \mathbb{Z}$, and

(6) the differentials

$$d_r(\mathcal{F}) : E_r^{s,}(\mathcal{F}) \longrightarrow E_r^{s+r,}(\mathcal{F})$$

are morphisms of $H^*(B; \mathbb{Z}_p)$-modules of degree $1-r$ and

$$E_{r+1}(\mathcal{F}) \cong H(E_r(\mathcal{F}); d_r(\mathcal{F}))$$

as graded-graded $H^*(B; \mathbb{Z}_p)$-modules;

(7) the convergence in (1) is as $H^*(B; \mathbb{Z}_p)$-modules, i.e., the filtration $F^q H^*(E; \mathbb{Z}_p)$ is by $H^*(B; \mathbb{Z}_p)$-submodules, and

$$E_0 H^*(E; \mathbb{Z}_p) \cong E_\infty(\mathcal{F})$$

as graded-graded $H^*(B; \mathbb{Z}_p)$-modules;

(8) under the identification (2) the $H^*(B; \mathbb{Z}_p)$-module structure on $E_2^{s,}(\mathcal{F})$ coincides with the standard $H^*(B; \mathbb{Z}_p)$-module structure of $\operatorname{Tor}^{s,*}_{H^*(B; \mathbb{Z}_p)} (H^*(Y; \mathbb{Z}_p), H^*(X; \mathbb{Z}_p))$.

These structures are compatible in the following sense;

(9) for each integer $r \geq 2$, the multiplication morphism

$$E_r^{s',*}(\mathcal{F}) \otimes_{\mathbb{Z}_p} E_r^{s'',*}(\mathcal{F}) \longrightarrow E_r^{s'+s'',*}(\mathcal{F})$$

and the comultiplication morphism

$$E_r^{s,*}(\mathcal{F}) \longrightarrow \bigoplus_{s'+s''=s} E_r^{s',*}(\mathcal{F}) \otimes_{\mathbb{Z}_p} E_r^{s'',*}(\mathcal{F})$$

are $H^*(B; \mathbb{Z}_p)$ balanced morphisms.

The spectral sequence, together with the super-structure detailed above is natural with respect to maps of 1-connected Hopf fibre squares of finite type. Moreover if $\varphi : \mathfrak{F}' \longrightarrow \mathfrak{F}''$ is a morphism of 1-connected Hopf fibre squares of finite type then the induced map of spectral sequences

$$\{E_r(\varphi)\} : \{E_r(\mathfrak{F}''), d_r(\mathfrak{F}'')\} \longrightarrow \{E_r(\mathfrak{F}'), d_r(\mathfrak{F}')\}$$

satisfies

(10) under the identification (2)

$$E_2(\varphi) = \mathrm{Tor}_{\varphi_B^*}(\varphi_X{}^*, \varphi_Y{}^*). \quad \square$$

This completes the description of the Eilenberg-Moore spectral sequence as we shall need it in our applications to Postnikov towers of Hopf spaces.

Our first task will be to provide means (i.e. tools) of computing with this spectral sequence. In the type of application that we invisage, we will be attempting to compute $H^*(E; \mathbb{Z}_p)$ from

$$H^*(X; \mathbb{Z}_p) \xleftarrow{f^*} H^*(B; \mathbb{Z}_p) \xrightarrow{g^*} H^*(Y; \mathbb{Z}_p).$$

Thus we find that our first order of business will be to compute

$$E_2(\mathfrak{F}) = \mathrm{Tor}_{H^*(B; \mathbb{Z}_p)}(H^*(X; \mathbb{Z}_p), H^*(Y; \mathbb{Z}_p)).$$

Now this is an algebraic problem. Namely, given morphism of commutative Hopf algebras of finite type

$$B \longleftarrow \Lambda \longrightarrow A$$

over $\mathbb{Z}_p$, compute the Hopf algebra $\mathrm{Tor}_\Lambda(B,A)$.

This is a viable problem, and we will "solve" it in the next section.

It is worthy of note that the analogous problem without the presence of Hopf algebra structures, i.e. , where $B \longleftarrow \Lambda \longrightarrow A$ are only commutative algebras and there morphisms seems many orders of magnitude more difficult. We will discuss this point somewhat in §8.

§ 2. Cohomology Hopf Algebras

Throughout this section k will denote a fixed field. Our algebraic terminology is taken from [3] and [30]. We will use upper index notation for our gradings. We assume familiarity with [30].

In this section we will study in earnest the problem of computing the algebra $\mathrm{Tor}_\Lambda(A,B)$ when we have been given k-algebra morphisms

$$A \xleftarrow{f} \Lambda \xrightarrow{g} B .$$

(Note that A is regarded as a right Λ-module and B as a left Λ-module via the structure morphisms

$$\varphi_A : A \otimes \Lambda \xrightarrow{1 \otimes f} A \otimes A \xrightarrow{\mu_A} A$$

$$\varphi_B : \Lambda \otimes B \xrightarrow{g \otimes 1} B \otimes B \xrightarrow{\mu_B} B ,$$

where as usual the unadorned $\otimes$ means $\otimes_k$.)

Now even in the case where A and Λ are polynomial algebras over k and B = k this seems to be a highly non-trivial problem. For while the Koszul resolution [7] [11] [39] in principal provides a computational tool it has proved impossible to evaluate the torsion product in closed form, or even to answer simple questions about the placement of the algebra generators. It is this placement of algebraic generators that has proved invaluable in the applications (see [7],[38],[39] etc.).

In this section we will find that the added presence of a Hopf algebra structure greatly simplifies the problem. We will only settle the result in a situation that we shall need later in this chapter, although the general case may be settled by the same techniques.

We turn first to the formulation of those results not quite to be found in [30] that we shall need. We will return to the study of Hopf algebras again in §6, to detail some further invariants needed for the solution we will present to the problem outlined in the introduction to this chapter.

Definition: A cohomology Hopf algebra is a connected positively graded Hopf algebra over k whose multiplication is commutative.

The category of cohomology Hopf algebras over k and their morphisms is denoted by $\mathcal{H}^*\mathcal{H}/k$.

The category $\mathcal{H}^*\mathcal{H}/k$ is pointed and k is the point. Indeed the category $\mathcal{H}^*\mathcal{H}/k$ is half-abelian, although we shall need little of these properties here. A thorough discussion may be found in [31].

Recollections: Let $\mathcal{A}$ be a pointed category with base point $*$. If $A', A'' \in \text{obj}\,\mathcal{A}$ then the composite $A' \longrightarrow * \longrightarrow A''$ is called the trivial morphism from A' to A".

Suppose that $f: A' \longrightarrow A''$ is a morphism in $\mathcal{A}$. A morphism $h : N \longrightarrow A'$ is called a kernel of f iff $f \circ h : N \longrightarrow A''$ is the trivial morphism, and whenever $g : B \longrightarrow A'$ is a morphism of $\mathcal{A}$ such that the composite $f \circ g : B \longrightarrow A'$ is trivial, there exists a unique morphism $\bar{g} : B \longrightarrow N$ so that the diagram

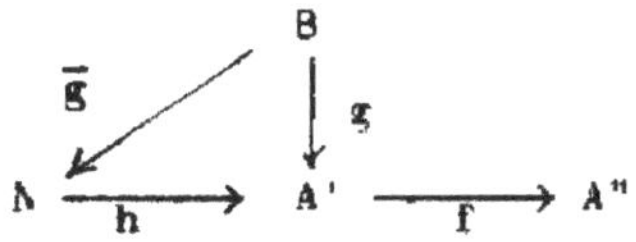

commutes. Cokernels are defined dually, namely a cokernel for f is a morphism $e: A'' \longrightarrow C$ such that $e \circ f : A' \longrightarrow C$ is the trivial morphism, and whenever $g : A'' \longrightarrow D$ is a morphism of $\mathcal{A}$ such that the composite $g \circ f : A' \longrightarrow C$ is trivial, there exists a unique morphism $\bar{g} : C \longrightarrow D$ so that the diagram

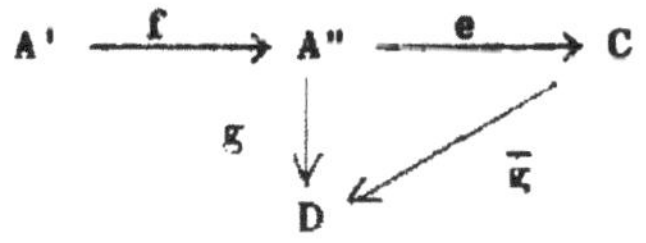

commutes.

<u>Proposition 2.1</u>: Let $A', A'' \in \text{obj } \mathcal{H}^*\mathcal{H}/k$ and $f : A' \longrightarrow A''$ a morphism of Hopf algebras. Then

(1) The natural morphism

$$e : A'' \longrightarrow k \times_{A'} A''$$

is a cokernel of f in $\mathcal{H}^*\mathcal{H}/k$, i.e. there is a natural cohomology Hopf algebra structure on $k \otimes_{A'} A''$ such that $e : A'' \longrightarrow k \otimes_{A'} A''$ is a morphism of Hopf algebras satisfying the universal property of a cokernel.

(2) If A' is abelian; i.e., the comultiplication in A' is also commutative, then the natural morphism

$$h : k \,\square_{A''}\, A' \longrightarrow A'$$

is a kernel of f in $\mathcal{H}^*\mathcal{H}/k$, i.e., there is a natural cohomology Hopf algebra structure on $k \,\square_{A''}\, A'$ such that $h : k \,\square_{A''}\, A' \longrightarrow A'$ is a morphism of Hopf algebras satisfying the universal property of a kernel. $\square$

<u>Notation</u>: If $f : A' \longrightarrow A''$ is a morphism of cohomology Hopf algebras then the cokernel of f, which exists and is given by 2.1 (1), will be denoted by $\operatorname{cok} f : A'' \longrightarrow A''/\!/f$.

If in addition A' is abelian, then 2.1 (2) assures us the existence of a kernel of f which we will denote by $\ker f : A' \backslash\!\backslash f \longrightarrow A'$.

Note that $\operatorname{cok} f : A'' \longrightarrow A''/\!/f$ is a surjection of the underlying k-modules, and $\ker f : A' \backslash\!\backslash f \longrightarrow A'$, when it exists, is an injec-tion of the underlying k-module structure.

Our objective in this section is to study the following:

<u>Problem</u>: Let $A', A'' \in \operatorname{obj} \mathcal{H}^*\mathcal{H}/k$ with A' abelian. Suppose that $f : A' \longrightarrow A''$ is a morphism of Hopf algebras. Compute $\operatorname{Tor}_{A'}(A'', k)$ where A'' is regarded as an A'-module via f and k via the augmentation.

Our main technical tool in studying this problem will be the following change of rings theorem. A proof may be found in [11; XVI. 6.1].

<u>Theorem 2.2</u>: Let A be a graded connected algebra over the field k. Let $B \subset A$ be a normal connected subalgebra and denote $k \otimes_B A$ by C. Suppose given an A-module N and

a C-module M. Finally assume that as a B-module A is projective. Then there exists a natural spectral sequence $\{E_r, d_r\}$ such that

$$E_r \Longrightarrow \mathrm{Tor}_A(M, N)$$

and

$$E_r^{p,q} = \mathrm{Tor}_C^p(M, \mathrm{Tor}_A^q(k, N)) .$$

(The differential d_r has bidegree (r, 1-r). There are auxillary gradings in Tor_(- , -) comming from the gradings of — and we have suppressed them from our notation.) □

In order to apply Theorem 2.2 to our problem we shall need the following classical result of Milnor and Moore [30; 4.4].

<u>Theorem 2.3 (Milnor-Moore)</u>: Let H' be a connected Hopf algebra over a Field k and $H'' \subset H'$ a Hopf subalgebra. Then H' is a free H"-module. □

Let us now consider how these results may be applied to our problem above. We will suppose given A', $A'' \in \mathcal{H}^*\mathcal{H}/k$ with A' abelian, and a morphism of Hopf algebras $f: A' \longrightarrow A''$. Let us denote by $C \subset A''$ the image of A' under f. Then C is certainly a sub-Hopf algebra of A". One may readily check that

$$C \cong k \otimes_{A' \mathbin{\backslash\!\backslash} f} A' .$$

Now clearly, $A' \mathbin{\backslash\!\backslash} f \hookrightarrow A'$ is a normal sub-Hopf algebra of A' and hence by Theorem 2.3 A' is a free $A' \mathbin{\backslash\!\backslash} f$ -module. Thus we may apply the change of rings theorem 2.2 to the A-module k and the C-module A". We obtain a spectral sequence $\{E_r, d_r\}$ with

$$E_r \Longrightarrow \mathrm{Tor}_{A'}(A'', k)$$

and

$$E_2^{p,q} = \mathrm{Tor}_C{}^p\,(A'', \mathrm{Tor}^q_{A'\backslash\backslash f}\,(k,k)).$$

Now as noted above $C \subset A''$ is a sub-Hopf algebra. Hence by a second application of Theorem 2.3 we find that A'' is a free C-module. Therefore

$$\mathrm{Tor}_C{}^p\,(A'', \mathrm{Tor}_{A'\backslash\backslash f}^q\,(k,k)) = 0$$

for $p < 0$.

Hence the above spectral sequence degenerates to the natural isomorphism

$$A'' \otimes_C \mathrm{Tor}_{A'\backslash\backslash f}(k,k) \cong \mathrm{Tor}_{A'}(A'',k)$$

given by the edge homomorphism.

Next we must note that as $A'\backslash\backslash f \subset A$ is central, A being commutative, the action of C on $\mathrm{Tor}_{A'\ \ f}(k,k)$ is trivial. Hence we find

$$A'' \otimes_C \mathrm{Tor}_{A'\backslash\backslash f}(k,k) \cong (A'' \otimes_C k) \otimes_k \mathrm{Tor}_{A'\backslash\backslash f}(k,k).$$

Quite obviously

$$A'' \otimes_C k \cong A'' \otimes_{A'} k\ .$$

Combining all this with the naturality of the spectral sequence 2.2 we find:

<u>Theorem 2.4</u> : Let $A', A'' \in \mathrm{obj}\,\mathcal{H}^*\mathcal{H}/k$ with A' abelian. Suppose that $f: A' \longrightarrow A''$ is a morphism of Hopf algebras. Then there is a natural isomorphism of Hopf algebras

$$\mathrm{Tor}_{A'}(A'',k) \cong A''/\!/f \otimes \mathrm{Tor}_{A'\backslash\backslash f}(k,k). \quad \square$$

Thus we have reduced our problem to the computation of $\mathrm{Tor}_H(k,k)$ when H is an abelian Hopf algebra over a field k. This we will do with the aid of the Borel structure theorem[30; 7.11] for commutative Hopf algebras over perfect field of nonzero characteristic.

Notation: For any graded vector space V over k denote by S[V] the symmetric algebra of V. Recall that this is the free commutative algebra generated by V. Suppose that k has characteristic $p \neq 0$. If $V^i = 0$ unless p^i is even we write P[V] for S[V]; it is just the polynomial algebra generated by V. If $V^i = 0$ unless p^i is odd then we write E[V] for S[V]; it is just the exterior algebra generated by V.

Finally if V is one dimensional, with basis vector x, we will often write P[x], S[x], and E[x] , in place of P[V], S[V], and E[V].

Recollections: Let H be a Hopf algebra over a field k and IH the augmentation ideal of H. The indecomposables of H, QH are defined by $QH = k \otimes_H IH$. Let JH be the coaugmentation ideal of H. The primitives of H, PH, are defined by $PH = JH \square_H k$. There is then a natural map of k-modules $\alpha_H : PH \longrightarrow QH$. If α_H is epic we say that H is primitive (or primitively generated). If α_H is monic we say that H is coprimitive.

Theorem 2.5 (Borel): Let H be an abelian connected Hopf algebra of finite type over a perfect field k of characteristic $p \neq 0$. Then H is isomorphic as an algebra to a tensor product of algebras of three basic types

E[x]	:	deg x = n,	pn odd,
P[x]	:	deg x = n,	pn even
$P[x]/(x^{p^r})$	:	deg x = n,	pn even .

If in addition H is primitively generated then H is isomorphic as a Hopf algebra to a tensor product of Hopf algebras of the above three basic types where each x is to be primitive. □

Let us return now to our problem. For our purposes it will suffice to consider

$$f : A' \longrightarrow A''$$

where A' is a primitively generated Hopf algebra, that as an algebra is free commutative. We shall find the following results useful.

<u>Proposition 2.6</u>: Let H be a Hopf algebra over a field k. Then

(1) if $H' \subset H$ is a sub-Hopf algebra and H is primitive so is H' ;

(2) if $H \longmapsto H''$ is a quotient Hopf algebra of H and H is coprimitive so is H".

<u>Proof</u>: If the characteristic of k is zero, this is instant from [30; 4.17]. So suppose that the characteristic of k is $p \neq 0$. Let

$$\xi_A : A \longrightarrow A \quad | \quad \xi(x) = \begin{cases} 0 \ , \text{ if } p(\deg x) \text{ is odd} \\ x^p \ , \text{ if } p(\deg x) \text{ is even.} \end{cases}$$

By [30;4.21] $\xi_H = 0$. Hence $\xi_{H''} = 0$ also. Applying [30; 4.21] a second time, this time to H" yields (2). The assertion (1) follows by considering the dual morphism λ [30; 4.22]. □

<u>Corollary 2.7</u>: Let H be a primitive Hopf algebra of finite type over a perfect field k of characteristic p.

Suppose that as an algebra H is free commutative. Then if $H' \subset H$ a Hopf subalgebra, H' is a primitive Hopf algebra of finite type that is isomorphic as an algebra to a free commutative algebra.

Proof: Immediate from the Borel structure theorem and 2.6. □

Thus in the situation of our problem

$$A' \xrightarrow[f]{} A''$$

where A' is primitive, and free commutative as an algebra. we find that $A'\backslash\backslash f$ is primitive and free commutative as an algebra. Thus in view of 2.4 we find that we must only discuss the structure of $\mathrm{Tor}_H(k,k)$ when H is a primitive Hopf algebra that is free commutative as an algebra. This we will do with the aid of the Koszul resolutions.

Let us recall that a free commutative algebra S[V] on a vector space V over k is always a tensor product

$$S[V] = P[V^+] \otimes_k E[V^-]$$

where

$$V^+ = \{V^n \mid pn \text{ even}\}$$

$$V^- = \{V^n \mid pn \text{ odd}\}$$

where p is the characteristic of k, $p \neq 0$. Since $\mathrm{Tor}_-(k,k)$ preserves tensor product it suffices to discuss the polynomial and exterior case separately.

Consider first the case $P[V^+]$, as it is easiest. We wish to compute the Hopf algebra $\mathrm{Tor}_{P[V^+]}(k,k)$. We do this by forming a nice resolution of k as a $P[V^+]$-module.

Notation: Let W be a graded vector space over k. Denote by $s^{m,n}(W)$ the bigraded vector space over k given by

$$(s^{m,n}(W))^{i,j} = \begin{cases} 0, & i \neq m \\ W^{n+j}_{,} & i = m. \end{cases}$$

We form the bigraded primitive Hopf algebra

$$\mathcal{E}(V^+) = P[V^+] \otimes E[\ s^{-1,0}\ V^+]$$

and define on it a derivation d of Hopf algebras by requiring

$$d : s^{-1,0}\ V^+ \longrightarrow V^+$$

be the natural isomorphism. One immediately checks

$$H(\mathcal{E} ; d) \cong k$$

as Hopf algebras over k. Thus

$$\mathrm{Tor}_{P[V^+]}(k,k) \cong H(\ k \otimes_{P[V^+]} \mathcal{E}(V^+);\ 1 \times d)$$
$$\cong E[\ s^{-1,0}\ V^+]$$

as bigraded Hopf algebras over k.

We turn now to the exterior algebra case. To this end we recall the definition of the Hopf algebra with divided powers generated by a vector space. Let p, the characteristic of k, be odd. Let W be a k-vector space concentrated in even degrees. We form a Hopf algebra $\Gamma[W]$ defined as follows. For every $w \epsilon W$ we are given elements

$$\gamma_i(w) \ \epsilon\ \Gamma[W] \ ; \quad \text{degree} \ \ \gamma_i(w) = i(\deg w)\ ,\ i>0$$
$$\gamma_1(w) = w\ .$$

These elements are to generate $\Gamma[W]$ subject only to the rules:

$$\gamma_i(w)\ \gamma_j(w) = \frac{(i+j)!}{i!\ \ j!}\ \gamma_{i+j}(w)$$
$$\gamma_i(u)\ \gamma_j(v) = \gamma_j(v)\ \gamma_i(u).$$

A comultiplication

$$\nabla : \Gamma[W] \longrightarrow \Gamma[W] \otimes \Gamma[W]$$

is defined by

$$\nabla \gamma_i(w) = \sum_{m+n=i} \gamma_m(w) \otimes \gamma_n(w)$$

where we use the convention $\gamma_0(w) = 1 \in \Gamma[W]^0$ for any $w \in W$. One may readily verify that $\Gamma[W]$ is a coprimitive Hopf algebra over k. Indeed $\Gamma[W]$ is the dual of the primitive Hopf algebra $P[W]$.

Note that as an algebra $\Gamma[W]$ is generated by the classes $\gamma_{p^r}(w)$, $w \in W$. More precisely

$$Q\Gamma[W] = \bigoplus_{r=0}^{\infty} \gamma_{p^r}(w).$$

This will be of importance in our later calculations. Note also that the primitive elements in $\Gamma[W]$ are just W, i.e.,

$$P\Gamma[W] = W.$$

Returning now to our study of

$$\mathrm{Tor}_{E[V^-]}(k,k)$$

we introduce the bigraded coprimitive Hopf algebra

$$\mathscr{E}(V^-) = E[V^-] \otimes \Gamma[\, s^{-1,0}\, V^+]$$

and define a derivation d of Hopf algebras by the requirement that

$$d(\, v \otimes \gamma_i(u)) = vu \otimes \gamma_{i-1}(u)$$

where $\gamma_{-1}(u) = 0$ for any $u \in s^{-1,0}V^+$.

One readily checks that $\mathscr{E}(V^-)$ is acyclic. (Indeed, ignoring the bigrading $\mathscr{E}(V^-)$ is dual to $\mathscr{E}(\, s^{-1,0}\, V^-)$.)

Thus as above we find

$$\mathrm{Tor}_{E[V^-]}(k,k) \cong \Gamma[\, s^{-1,0}\; V^-]$$

as bigraded Hopf algebras.

Summarizing we find:

Theorem 2.8: Suppose that k is a perfect field of non-zero characteristic. Let A', A" ε obj $\mathcal{H}^*\mathcal{H}/k$ and $f : A' \longrightarrow A"$ a morphism of Hopf algebras. Suppose that A' is a primitive (hence abelian) Hopf algebra, that as an algebra is isomorphic to a free commutative algebra. Then there is a natural isomorphism

$$\mathrm{Tor}_{A'}(A",k) \cong A"//f \otimes E[s^{-1,0}QA' \setminus f^+] \otimes \Gamma[s^{-1,0}QA' \setminus f^-]$$

of bigraded Hopf algebras over k. □

3. Cohomology of Hopf Fibre Squares : E_2 and Beyond

We turn now to the task of computing the cohomology relations in Hopf fibre squares that we shall need for our intended study of a generalized Cartan-Serre theorem. For a 1-connected Hopf fibre square we will find that the structure theorem 2.8 will allow us to compute the E_2 term of the $\mathbb{Z}_p$-cohomology spectral sequence of $\mathcal{F}$ in many cases of interest. To proceed further we will find it convenient to divide our work into several stages. The case p = 2 is somewhat degenerate and is easily disposed of. In the case $p > 2$ we will find that there is one differential that must be computed. This we will do by the method of universal example in the following sections. After developing some necessary Hopf algebra techniques in the ensuing sections the stage will then be set for our generalized Cartan-Serre theorem.

We will suppose given a 1-connected Hopf fibre square

$$\mathfrak{F}\qquad \begin{array}{ccc} E & \longrightarrow & Y \\ \downarrow & & \downarrow g \\ X & \xrightarrow[f]{} & B \end{array}$$

of finite type, and a prime number p, satisfying the following conditions:

(1) $g: Y \longrightarrow B$ is the path space fibration over B;

(2) as an algebra, $H^*(B, \mathbb{Z}_p)$ is isomorphic to a free commutative algebra;

(3) $H^*(B; \mathbb{Z}_p)$ is an abelian Hopf algebra.

Notation and Conventions: Throughout the remainder of this section $\mathfrak{F}$ will denote a 1-connected Hopf fibre square of finite type satisfying the above restrictions for the prime p.

All cohomology in this section will be taken with respect to $\mathbb{Z}_p$, where p is the prime number fixed above. We will allow ourselves to simplify our notation by writing $H^*(\ )$ instead of $H^*(\ ;\mathbb{Z}_p)$.

Remarks: (1) Suppose that A is an algebra over the Steenrod algebra $\mathcal{A}^*(p)$ [27]. The global definition of the indecomposable elements QA of A together with the Cartan-formula shows that QA is an $\mathcal{A}^*(p)$-module.

More precisely if we denote by $\mathcal{U}\mathcal{A}/\mathcal{A}^*(p)$ the category of unstable algebras over the Steenrod algebra and $\mathcal{U}\mathcal{M}/\mathcal{A}^*(p)$ the category of unstable modules over the Steenrod algebra then Q is a functor

$$Q:\ \mathcal{U}\mathcal{A}/\mathcal{A}^*(p) \longrightarrow \mathcal{U}\mathcal{M}/\mathcal{A}^*(p).$$

The adjoint of this functor is the free unstable algebra construction of [27], [49].

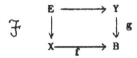

(2) Suppose that H' and H" are Hopf algebras over the Steenrod algebra (see for example [42] for the relevant definitions) and

$$f : H' \longrightarrow H''$$

is a morphism of Hopf algebras over $\mathcal{A}^*(p)$. The global definition of $H' \backslash\backslash f$, when it exists, shows that it too is a Hopf algebra over $\mathcal{A}^*(p)$. If H' is unstable so is $H' \backslash\backslash f$.

In particular if

$$f : X'' \longrightarrow X'$$

is a morphism of Hopf spaces, and $H^*(X') \backslash\backslash f^*$ is defined, then it is a Hopf algebra over the Steenrod algebra.

We will make free use of these remarks in the sequel. They are quite useful.

With the above conventions and remarks in mind we obtain in view of 2.8 the following result.

<u>Proposition 3.1</u>: Let $\{E_r(\mathfrak{F}); d_r(\mathfrak{F})\}$ denote the $\mathbb{Z}_p$-cohomology spectral sequence of the Hopf fibre square $\mathfrak{F}$. Then

$$E_2(\mathfrak{F}) = H^*(X) /\!/ f^* \otimes E[s^{-1,0}QH^*(B) \backslash\backslash f^{*+}] \otimes [s^{-1,0}QH^*(B) \backslash\backslash f^{*-}]$$

as Hopf algebras over the Steenrod algebra $\mathcal{A}^*(p)$.$\square$

At this point the cases p = 2 and p > 2 separate, the case p = 2 being degenerate. For when p = 2

$$QH^*(B) \backslash\backslash f^{*+} = QH^*(B) \backslash\backslash f^*$$

$$QH^*(B) \backslash\backslash f^{*-} = 0.$$

Thus we find:

<u>Theorem 3.2</u>: Let $\{E_r(\mathfrak{F}), d_r(\mathfrak{F})\}$ denote the $\mathbb{Z}_2$- cohomology spectral sequence of the Hopf fibre square $\mathfrak{F}$. Then $E_2(\mathfrak{F}) = E_\infty(\mathfrak{F})$ and there is a filtration

$\{F^qH^*(E)\}$ such that

$$E_0H^*(E) \cong H^*(X)//f^* \otimes E[s^{-1,0}QH^*(B)\backslash\backslash f^*]$$

as bigraded Hopf algebras over the Steenrod algebra $\mathcal{A}^*(2)$.

Proof: According to 3.1

$$E_2(\mathfrak{F}) \cong H^*(X)//f^* \otimes E[s^{-1,0}QH^*(B)\backslash\backslash f^*]$$

as Hopf algebras over the Steenrod algebra $\mathcal{A}^*(2)$, by our above remark concerning the special features of 3,1(and/or 2.8) when $p = 2$.

Thus we find that as an algebra over $\mathbb{Z}_2$, $E_2(\mathfrak{F})$ is generated by

$$H^*(X)//f^* = E_2^{-1,*}(\mathfrak{F})$$

and

$$s^{-1,0}QH^*(B)\backslash\backslash f^* \subset E_2^{-1,*}(\mathfrak{F}).$$

Recall that

$$d_r(\mathfrak{F}) : E_r^{s,t}(\mathfrak{F}) \longrightarrow E_r^{s+r,t+1-r}(\mathfrak{F}),$$

and that

$$E_r^{s,t}(\mathfrak{F}) = 0 \qquad \text{for } s > 0.$$

Thus, inductively, starting with $r = 2$ we find

$$d_r(\mathfrak{F}) : E_2^{0,*}(\mathfrak{F}) \longrightarrow E_2^{r,*}(\mathfrak{F}) = 0$$

and

$$d_r(\mathfrak{F}) : E_2^{-1,*}(\mathfrak{F}) \longrightarrow E_2^{r-1,*}(\mathfrak{F}) = 0.$$

Hence, inductively, $d_r(\mathfrak{F})$ must vanish on the algebra generators of $E_2(\mathfrak{F})$. Being a derivation it must therefore vanish. Thus, inductively from $r = 2$ we find

$$d_r(\mathfrak{F}) = 0 \quad \text{for all } r \geq 2.$$

Hence $E_2(\mathfrak{F}) = E_\infty(\mathfrak{F})$. The result now follows from 1.1. □

Now nothing like the above argument may be applied to

the case $p > 2$. Indeed we shall find that there is a non-zero differential in this case. We turn to this now.

<u>Convention</u>: Untill further notice we will assume that the prime p is > 2.

<u>Proposition 3.3</u>: Let $\{E_r(\mathfrak{F});\ d_r(\mathfrak{F})\}$ denote the $\mathbb{Z}_p$-cohomology spectral sequence of the Hopf fibre square $\mathfrak{F}$. Then the first possible non-zero differential $d_r(\mathfrak{F})$, $r \geq 2$, is $d_{p-1}(\mathfrak{F})$.

<u>Proof</u>: This will follow from an "accounting" (= bookkeeping) argument based on the whereabouts of the indecomposable and primitive elements in $E_2(\mathfrak{F})$. We will of course be using the fact that $d_r(\mathfrak{F})$ is a derivation of Hopf algebras.

Let us first study where the generators of $E_2(\mathfrak{F})$ lie. From 3.1 and our remarks in the previous section concerning $E[W]$ and $\Gamma[W]$ we find that $E_2(\mathfrak{F})$ is generated by

$$H^*(X) /\!/ f^* = E_2^{0,*}(\mathfrak{F})$$

$$s^{-1,0}\, QH^*(B) \backslash\!\backslash f^* \subset E_2^{-1,*}(\mathfrak{F})$$

and

$$\gamma_{p^t}\, s^{-1,0}\, QH^*(B) \backslash\!\backslash f^{*-} \subset E_2^{-p^t,*}(\mathfrak{F}) \qquad \text{for } t > 1$$

as an algebra over $\mathbb{Z}_p$. Thus in the usual planer diagram for $E_2(\mathfrak{F})$ the generators lie on the indicated vertical lines in the diagram below.

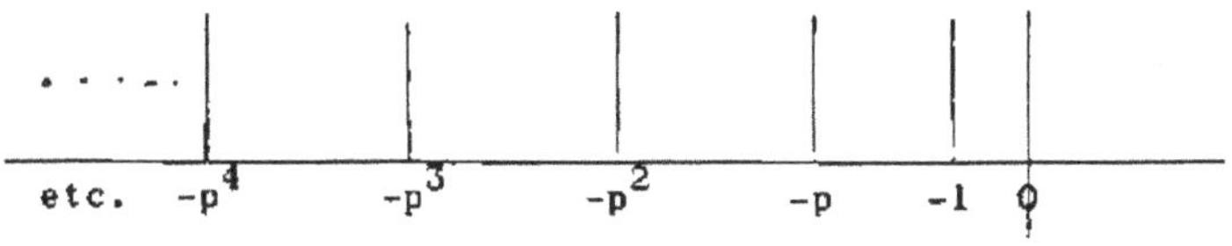

<u>Location of algebra generators for $E_2(\mathfrak{F})$.</u>

Let us next look for the primitive elements in $E_2(\mathfrak{F})$. As the primitive functor commutes with tensor products we find that the primitive elements of $E_2(\mathfrak{F})$ are given by

$$PH^*(X)/\!/f^* \subset E_2^{0,*}(\mathfrak{F})$$

and

$$s^{-1,0}\, QH^*(B)\backslash\!\backslash f^* \subset E_2^{-1,*}(\mathfrak{F}).$$

Now our proposition will follow from the following:

<u>Lemma</u>: Let $\{H, d\}$ be a differential Hopf algebra over a field k. Suppose that

$$x \in H$$

is an element of minimal degree such that

$$dx \neq 0 \in H.$$

Then x is indecomposable and dx is primitive.

<u>Proof</u>: Since d is a derivation

$$d(uv) = (-1)^{\varepsilon}\, u\, dv + du\, v\,.$$

Thus if

$$x = u\, v \qquad \deg u, \deg v > 0.$$

Then

$$du \neq 0 \quad \text{or} \quad dv \neq 0$$

contrary to the minimality of degree x. Therefore

$$x \neq u\, v \quad \deg u, \deg v > 0,$$

and hence x is indecomposable. Next note that

$$\Delta(x) = x \otimes 1 + 1 \otimes x + \sum_{\substack{\deg x_i' \\ \deg x_i''}\Big\} < \deg x} x_i' \otimes x_i''$$

where

$$\Delta : H \longrightarrow H \otimes H$$

is the comultiplication in H. The differential d being

compatible with the coalgebra structure yields

$$\Delta(dx) = dx \otimes 1 + 1 \otimes dx + \sum_{\substack{\deg x_i' \\ \deg x_i''}\} < \deg x} dx_i' \otimes x_i'' + (-1)^{\varepsilon_i} x_i' \otimes dx_i''$$

$$= dx \otimes 1 + 1 \otimes dx$$

as $dx_i' = 0 = dx_i''$ since $\deg x_i'$, $\deg x_i'' < \deg x$.

Thus dx is indeed primitive. **

Now let us suppose that $d_r(\mathfrak{F})$ is the first non-zero differential with $r \geq 2$. Then $E_2(\mathfrak{F}) = E_r(\mathfrak{F})$. According to the lemma there then exists an indecomposable element

$$x \in E_2(\mathfrak{F})$$

such that

$$d_r(\mathfrak{F})(x) \neq 0 \in E_2(\mathfrak{F})$$

and $d_r(\mathfrak{F})(x)$ is primitive. Thus from our discussion preceeding the lemma

$$\deg x = (-p^t, s) \quad \text{or } (0,s)$$

and

$$\deg d_r(\mathfrak{F})(x) = (-1, q) \quad \text{or } (o, q).$$

We also compute

$$\deg d_r(\mathfrak{F})(x) = (-p^t + r, s + 1 - r) \text{ or } (r, s+1-r).$$

As $E_2^{r,*}(\mathfrak{F}) = 0$ we find we must have :

$$(-p^t+r, s+1-r) = (-1,q) \quad \text{or } (o,q)$$

by equating the two computations of degree $d_r(\mathfrak{F})(x)$ and elimination of the possibility $\deg x = (0,s)$; for then $d_r(\mathfrak{F})(x)$ must be zero. Clearly the smallest integer $r \geq 2$ for which either of these equalities can hold is $r = p-1$ when $t = 1$. □

Thus we find that our next step will have to be the

study of the differential $d_{p-1}(\mathfrak{F})$.

§4. Cohomology of Hopf Fibre Squares: A Differential

Let us continue to suppose that $\mathfrak{F}$ is a Hopf fibrē square and p an odd prime satisfying the restrictions of the previous section. We have seen how to compute $E_{p-1}(\mathfrak{F})$. Our next task is to compute $d_{p-1}(\mathfrak{F})$ in the hopes of gaining more information on $H^*(E)$. If we adopt the point of view that $d_{p-1}(\mathfrak{F})$ is a sort of cohomology operation, the method of universal examples suggests itself as a possible means of extracting a formula for $d_{p-1}(\mathfrak{F})$. Here is how the method may be applied.

Suppose that we choose an element

$$x \in QH^*(B)\backslash\backslash f^{*-}$$

and wish to compute

$$d_{p-1}(\mathfrak{F})\, \gamma_p(s^{-1,0}x) \in QH^*(B)\backslash\backslash f^*.$$

Let us first choose a class $\bar{x} \in H^*(B)\backslash\backslash f^* \subset H^*(B)$ that represents x. As $\bar{x}$ is a $\mathbb{Z}_p$- cohomology class we may then find a map

$$\varphi : B \longrightarrow K(\mathbb{Z}_p, 2t+1)$$

where

$$\deg x = 2t+1$$

that represents $\bar{x}$, i.e.

$$\varphi^* \iota_{2t+1} = \bar{x}$$

where

$$\iota_{2t+1} \in H^{2t+1}(K(\mathbb{Z}_p, 2t+1))$$

is the fundamental class.

<u>Proposition 4 1</u>: We may choose

$$\varphi : B \longrightarrow K(\mathbb{Z}_p, 2t+1)$$

to be a morphism of Hopf spaces.

<u>Proof</u>: We wish to show that we may arrange things so that the diagram

$$\begin{array}{ccc} B \times B & \xrightarrow{\ \mu\ } & B \\ {\scriptstyle \varphi \times \varphi}\downarrow & & \downarrow{\scriptstyle \varphi} \\ K \times K & \xrightarrow{\ \mu\ } & K \end{array}$$

is homotopy commutative, where we have abbreviated $K(\mathbb{Z}_p, 2t+1)$ to K.

Now the two mappings

$$\varphi \cdot \mu \ , \mu(\varphi \times \varphi) : B \times B \longrightarrow K$$

represent cohomology classes

$$a, b \in H^{2t+1}(B \times B; \mathbb{Z}_p).$$

From the universal property of K, i.e., that it represents the functor $H^{2t+1}(\ \ ; \mathbb{Z}_p)$ we find that the desired diagram commutes iff

$$a = b \in H^{2t+1}(B \times B; \mathbb{Z}_p).$$

Now it is an easy matter to compute these two cohomology classes. For by the Künneth theorem

$$H^{2t+1}(B \times B; \mathbb{Z}_p) = [H^*(B; \mathbb{Z}_p) \otimes H^*(B; \mathbb{Z}_p)]^{2t+1}$$

and as K is 2t-connected

$$H^{2t+1}(K \times K; \mathbb{Z}_p) = H^{2t+1}(K; \mathbb{Z}_p) \otimes \mathbb{Z}_p + \mathbb{Z}_p \otimes H^{2t+1}(K; \mathbb{Z}_p)$$

Thus

$$a = \mu^* \varphi^*(\iota) = \mu^*(\bar{x})$$

while

$$b = (\varphi \times \varphi)^* \mu^*(\iota) = (\varphi \times \varphi)^*(\iota \otimes 1 + 1 \otimes \iota)$$
$$= \bar{x} \otimes 1 + 1 \otimes \bar{x}.$$

Thus we find that it suffices to show that $\bar{x} \in H^{2t+1}(B;\mathbb{Z}_p)$ is a primitive element.

But consider, $H^*(B;\mathbb{Z}_p)$ is an abelian Hopf algebra over $\mathbb{Z}_p$. By Borel's structure theorem applied to $H^*(B;\mathbb{Z}_p)$ and its dual [30;7.11] we therefore find an isomorphism of Hopf algebras

$$H^*(B;\mathbb{Z}_p) \cong E[V] \otimes H$$

where V is concentrated in odd degrees and H in even degrees. Hence all the odd dimensional indecomposables of $H^*(B;\mathbb{Z}_p)$ are primitive which now yields the desired conclusion. (Remark: Clearly it would suffice to show that in the dual Hopf algebra of $H^*(B;\mathbb{Z}_p)$ all the odd dimensional primitives are indecomposable. But as this dual has a commutative multiplication [30; §4] yields this up instantly.)□

Thus certainly the composite

$$X \xrightarrow{f} B \xrightarrow{\varphi} K(\mathbb{Z}_p, 2t+1)$$

is a morphism of Hopf spaces.

Let us denote by

$$L(\mathbb{Z}_p, 2t+1) \longrightarrow K(\mathbb{Z}_p, 2t+1)$$

the path space fibration over $K(\mathbb{Z}_p, 2t+1)$. We may then form the Hopf fibre square

$$\mathfrak{F}_x \qquad \begin{array}{ccc} W & \longrightarrow & L(\mathbb{Z}_p, 2t+1) \\ \downarrow & & \downarrow \\ X & \xrightarrow{\varphi f} & K(\mathbb{Z}_p, 2t+1). \end{array}$$

Clearly the diagram

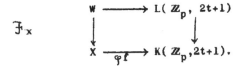

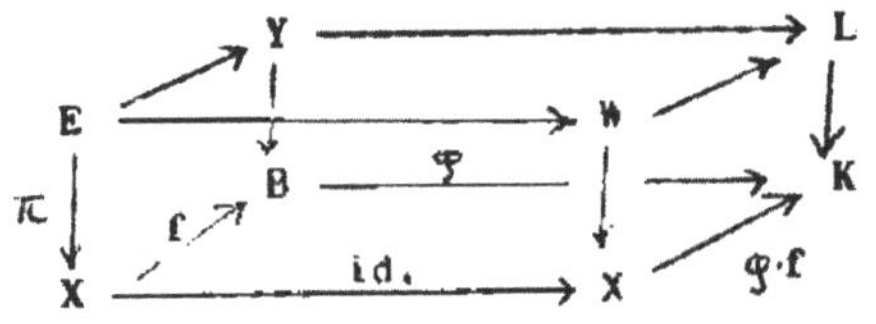

is clearly a morphism of Hopf squares

$$\psi : \mathfrak{F} \longrightarrow \mathfrak{F}_X$$

Let us examine now the $\mathbb{Z}_p$-cohomology spectral sequence of the Hopf fibre square $\mathfrak{F}_X$. Clearly we will wish to apply 3.1 again. For this application we need only recall that as $K(\mathbb{Z}_p,n)$ is an infinite loop space, $H^*(K(\mathbb{Z}_p,n))$ is certainly abelian, while Cartan [10] has shown that $H^*(K(\mathbb{Z}_p,n))$ is always a free commutative algebra for $n > 1$. (We shall presently need a more precise knowledge of the $\mathbb{Z}_p$-cohomology of $K(\mathbb{Z}_p,n)$ and we have therefore included a more precise statement of Cartan's results in the next section.) Now a certain simplification in applying 3.1 results from observing that

$$f^* \varphi^* : H^*(K(\mathbb{Z}_p,2t+1)) \longrightarrow H^*(X)$$

is the zero map. For, by the choice of $\bar{x}$ as an element of $H^*(B)\backslash\backslash f^*$ we clearly have

$$f^* \varphi^*(\iota) = f^* \bar{x} = 0 \in H^{2t+1}(X).$$

As $K(\mathbb{Z}_p,2t+1)$ represents the functor $H^{2t+1}(\ ; \mathbb{Z}_p)$ we must have

$$\varphi f : X \longrightarrow K(\mathbb{Z}_p, 2t+1)$$

is null-homotopic, and hence certainly induces the zero map in $\mathbb{Z}_p$-cohomology. Thus we have:

<u>Proposition 4.2</u>: Let $\{E_r(\mathfrak{F}_X); d_r(\mathfrak{F}_X)\}$ denote the $\mathbb{Z}_p$-cohomology spectral sequence of the Hopf fibre square $\mathfrak{F}_X$. Then

$$E_2(\mathfrak{F}_X) = H^*(X) \otimes E[s^{-1,0}QH^*(K)^+] \otimes \Gamma[s^{-1,0}QH^*(K)^-],$$

where as before $K = K(\mathbb{Z}_p, 2t+1)$. □

Let us now examine the map

$$E_2(\psi) : E_2(\mathcal{F}_X) \longrightarrow E_2(\mathcal{F}).$$

For this we will need the following remarks.

Recollections: Let $\mathcal{M}/k$ denote the category of moduled over the field k and $\mathcal{A}/k$ the category of graded connected k-algebras. Then

$$Q : \mathcal{A}/k \rightsquigarrow \mathcal{M}/k$$

is a functor.

The Hopf algebra constructions, E[ ], [ ] and P[ ] of the previous section yield functors

$$E, \Gamma, P : \mathcal{M}/k \longrightarrow \mathcal{H}^*\mathcal{H}/k.$$

Proposition 4.3: The mapping

$$\psi : \mathcal{F} \longrightarrow \mathcal{F}_X$$

induces the map

$$E_2(\psi) = \eta \otimes E[\, s^{-1,0}Qf^{*+}] \otimes \Gamma[\, s^{-1,0}Qf^{*\,-}]$$

of the $\mathbb{Z}_p$-cohomology spectral sequences of the fibre squares $\mathcal{F}$ and $\mathcal{F}_X$, where

$$\eta : H^*(X) \longrightarrow H^*(X)//f^*$$

is the natural quotient mapping.

Proof: This is immediately obvious by using the Koszul complex constructions of the last section. □

In particular we find that

$$\begin{aligned} E_2(\psi)(s^{-1,0}\gamma_{p^t}(\iota)) &= s^{-1,0}\gamma_{p^t}(\varphi^*(\iota)) \\ &= s^{-1,0}\gamma_{p^t}(x) \end{aligned}$$

Thus by naturality

$$d_{p-1}(\mathcal{F})(s^{-1,0}\gamma_{p^t}(x)) = E_2(\psi)d_{p-1}(\mathcal{F}_X)(s^{-1,0}\gamma_{p^t}(\iota)).$$

So it will suffice for us to study $\mathcal{F}_X$. This study is simplified even more as follows.

<u>Proposition 4.4</u>: Let

$$\mathcal{L}(2t+1) \qquad \begin{array}{ccc} K(\mathbb{Z}_p, 2t) & \longrightarrow & L(\mathbb{Z}_p, 2t+1) \\ \downarrow & & \downarrow \\ * & \longrightarrow & K(\mathbb{Z}_p, 2t+1) \end{array}$$

be the Hopf fibre square formed by pullback over a point. Then there is a natural isomorphism of spectral sequences

$$\{E_r(\mathcal{F}_X), d_r(\mathcal{F}_X)\} \cong \{H^*(B) \otimes E_r(\mathcal{L}(2t+1), 1 \otimes d_r(\mathcal{L}(2t+1))\}.$$

<u>Proof</u>: This is an immediate consequence of functorality and the fact that

$$W \cong B \times K(\mathbb{Z}_p, 2t)$$

since

$$\wp f : B \longrightarrow K(\mathbb{Z}_p, 2t+1)$$

is null homotopic. □

Thus we find that we must only study the spectral sequence

$$\{E_r(\mathcal{L}(2t+1)); d_r(\mathcal{L}(2t+1))\}.$$

To this end we recall a few facts:

<u>Recollections</u>: Let X be a simply connected space and $PX \longrightarrow X$ the path space fibration. We may form a fibre square

$$\mathcal{L}(X) \qquad \begin{array}{ccc} \Omega X & \longrightarrow & PX \\ \downarrow & & \downarrow \\ * & \longrightarrow & X \end{array}$$

by pullback over a point $* \in X$. The $\mathbb{Z}_p$-cohomology spectral sequence $\{E_r(\mathcal{L}), d_r(\mathcal{L})\}$ of $\mathcal{L}$ then we have

$$E_2(\mathcal{L}) = \mathrm{Tor}_{H^*(X)}(\mathbb{Z}_p, \mathbb{Z}_p).$$

As is well known [3] [11]

$$E_2^{0,*}(\mathcal{L}) = \mathbb{Z}_p$$

and

$$E_2^{-1,*}(\mathcal{L}) = s^{-1,0}QH^*(X).$$

Since

$$d_r(\mathcal{L})(E_2^{-1,*}(\mathcal{L})) \subset E_2^{r-1,*}(\mathcal{L}) = 0 \qquad \text{for } r\geq 2,$$

it follows that the elements

$$s^{-1,0}QH^*(X) = E_2^{-1,*}(\mathcal{L})$$

are infinite cycles. There is thus defined a sort of edge mapping

$$s^{-1,0}QH^*(X) \longrightarrow F^{-1}H^*(\Omega X; \mathbb{Z}_p)/\ F^0H^*(\Omega X; \mathbb{Z}_p)$$
$$\subset \widetilde{H}^*(\Omega X; \mathbb{Z}_p).$$

The composite

$$s : QH^*(X) \longrightarrow \widetilde{H}^*(\Omega X; \mathbb{Z}_p)$$

then coincides with the classical cohomology suspension mapping. [39] ([10] [12] [13] for the classical cohomology suspension.)

We may now complete our work in this section with the aid of :

<u>Proposition 4.5 (Cartan)</u>: The element

$$\beta P_p^t \iota \in H^{2tp+2}(K(\mathbb{Z}_p, 2t+1))$$

is an indecomposable element in the kernel of the cohomology suspension

$$s: QH^*(K(\mathbb{Z}_p, 2t+1)) \longrightarrow H^*(K(\mathbb{Z}_p, 2t)).$$

<u>Proof</u>: See [10; Expose16; 2.1]. ☐

Proposition 4.6 : Let $\{E_r(2t+1); d_r(2t+1)\}$ denote the $\mathbb{Z}_p$-cohomology spectral sequence of the fibre square

$$\mathcal{L}(2t+1) \qquad \begin{array}{ccc} K(\mathbb{Z}_p, 2t) & \longrightarrow & L(\mathbb{Z}_p, 2t+1) \\ \downarrow & & \downarrow \\ * & \longrightarrow & K(\mathbb{Z}_p, 2t+1). \end{array}$$

Then

$$d_{p-1}(2t+1)(\gamma_p(s^{-1,0}\iota)) = \lambda s^{-1,0}\beta P_p^t \iota \in E_2^{-1,*}(2t+1)$$

where $\lambda \neq 0 \in \mathbb{Z}_p$. Thus more generally we have

$$d_{p-1}(2t+1)\ \gamma_j(s^{-1,0}\iota) = (\lambda s^{-1,0}\ \beta P_p^t \iota)(\gamma_{j-1}(s^{-1,0}\ \iota))$$

for all $j \geq p$.

Proof: This is just a combination of previous results coupled with careful bookkeeping .

From Proposition 4.5 it follows that

$$s^{-1,0}\beta P_p^t \iota \ \in E_2^{-1,*}(2t+1)$$

must be a boundary under some differential in the spectral sequence. According to 3.3 the first possible differential that can do the trick is $d_{p-1}(2t+1)$. The following lemma shows that it is the only one.

Lemma: Let i be a positive integer. Then

$$E_2^{-i,j}(2t+1) = 0$$

unless $j \geq i(2t+1)$.

Proof: By results of Cartan

$$H^*(K(\mathbb{Z}_p, 2t+1))$$

is a free commutative algebra. By definition $K(\mathbb{Z}_p, 2t+1)$ is 2t-connected. Now apply the Koszul complex construction to compute

$$\mathrm{Tor}_{H^*(K(\mathbb{Z}_p, 2t+1))}(\mathbb{Z}_p , \mathbb{Z}_p)$$

keeping track of degrees. **

Now suppose that

$$s^{-1,0}\beta P_p^t\iota \quad \varepsilon\ E_2^{-1,\ 2tp+2}\ (2t+1)$$

were to survive to $E_r(2t+1)$ and is a boundary under $d_r(2t+1)$. Thus

$$s^{-1,0}\beta P_p^t\iota \ = \ d_r(2t+1)(u)$$

for some $u\ \varepsilon\ E_r^{-r-1,2tp+2+r-1}(2t+1)$. By the lemma we would then have

$$2tp+2+r-1 \geq (r+1)(2t+1) \ = 2tr+r+2t+1$$

which yields

$$p \geq r+1$$

or

$$r \leq p-1.$$

Hence the only possible way

$$s^{-1,0}\beta P_p^t\iota \ \ \varepsilon\ \ E_2^{-1,\ 2tp+2}(2t+1)$$

can be a non-boundary is for

$$s^{-1,0}\beta P_p^t \quad = d_{p-1}(2t+1)(u)$$

for some

$$u\ \varepsilon\ E_2^{-p,\ 2tp+p}(2t+1)\ .$$

But

$$\gamma_p(s^{-1,0}\ \iota\ )\ \varepsilon\ E_2^{-p,\ 2tp+p}(2t+1)$$

and is a $\mathbb{Z}_p$-basis for this $\mathbb{Z}_p$-vector spce, whence the formula

$$d_{p-1}(2t+1)\,\gamma_p(s^{-1,0}\iota) \ = \lambda s^{-1,0}\beta P_p^t\iota$$

follows.

The remaining formula follow instantly by using the compatibility of the coalgebra structure with the differential. □

Summing up we can state:

Theorem 4.7: Let $\{E_r(\mathfrak{F}), d_r(\mathfrak{F})\}$ denote the $\mathbb{Z}_p$-cohomology spectral sequence of the Hopf fibre square $\mathfrak{F}$. Then for p an odd prime

$$E_2(\mathfrak{F}) \cong H^*(X) /\!/ f^* \otimes E[s^{-1,0}QH^*(B) \backslash\backslash f^{*+}] \otimes \Gamma[s^{-1,0}QH^*(B) \backslash\backslash f^{*-}]$$

as Hopf algebras over the Steenrod algebra $\mathcal{A}^*(p)$.

Moreover the first non-zero differential is $d_{p-1}(\mathfrak{F})$ which is given by the formula (on algebra generators)

$$d_{p-1}(\mathfrak{F})(\gamma_{p^q}(s^{-1,0}x)) = (\lambda s^{-1,0}\beta P^t_p x)(\gamma_{p^q-1}(s^{-1,0}x))$$

for each $q > 0$, where $\lambda \neq 0 \in \mathbb{Z}_p$, $x \in QH^*(B) \backslash\backslash f^{*-}$.

Proof: The only thing remaining to be proved is the formula for $d_{p-1}(\mathfrak{F})$. However this follows from 4.2, 4.3 and 4.6. □

Thus we find that for p-odd, we must know a little more of the action of the Steenrod algebra on $H^*(B)$ before we can proceed further. Of course on the basis of 4.7 we may compute $E_p(\mathfrak{F})$. However there may be many more non-zero differentials (see e.g. [31]).

However the formula 4.7 is extremely useful and can be applied with extreme effectiveness in many cases of interest (see e.g. [31] [37] [38] [40] [41] and the next section.) One such case is that of the stages in a stable Postnikov tower and we turn to this in the next section.

§5 Stable Postnikov Systems

We turn now to specializing the results of the previous sections to the case of stable Postnikov systems. We begin as usual with a few definitions.

Definition: Let A be a graded abelian group of finite type. We denote by K(A) the space

$$\bigtimes_{n=0}^{\infty} K(A_n, n)$$

and refer to it as the Eilenberg-MacLane space of type A. The path space fibration over KA is denoted $LA \longrightarrow KA$.

The space K(A) inherits a natural Hopf space structure from the operation of addition in A. Here is one way to describe it.

Note that the Hurecwicz map

$$h: \pi_*(KA) \longrightarrow H_*(K(A); \mathbb{Z})$$

is a monomorphism onto a direct summand. We also have

$$h: \pi_*(KA \times KA) \longrightarrow H_*(K(A) \times K(A); \mathbb{Z})$$

is monic onto a direct summand. Thus since

$$\pi_*(KA \times KA) \cong A \times A$$

we obtain

$$\mu : H_*(KA \times KA; \mathbb{Z}) \longrightarrow A \times A \xrightarrow{\alpha} A$$

where α is addition. This composite is a cohomology class

$$\mu \in H^*(KA \times KA; A)$$

which, since KA represents $H^*(- ;A)$ is realized by a unique homotopy class

$$\mu : KA \times KA \longrightarrow KA$$

which is a Hopf space structure on KA as may easily be checked.

Convention: While in general the space KA carries many inequivalent Hopf space structures [25][31,II] we will only be concerned with the one described above.

Definition: A stable r-stage Postnikov system is defined inductively as follows

(1) A stable 1-stage Postnikov system is an Eilenberg-MacLane space $K(A)$, where A is a positively graded abelian group; equipped with its natural Hopf space structure.

(2) For $r > 1$, a stable r-stage Postnikov system is a pair consisting of a Hopf space E, and a Hopf fibre square

$$\mathfrak{F}(E) \qquad \begin{array}{ccc} E & \longrightarrow & L(A) \\ \downarrow & & \downarrow \\ X & \longrightarrow & K(A) \end{array}$$

where A is a 1-connected graded abelian group of finite type and X is a stable r-1 stage Postnikov system.

Abuse of Terminology: Following the usual practice of abusing terminology, we shall say that a Hopf space E is a stable r-stage Postnikov system when there exists a stable r-stage Postnikov system.

$$\mathfrak{F} \qquad \begin{array}{ccc} E & \longrightarrow & L(A) \\ \downarrow & & \downarrow \\ B & \longrightarrow & K(A) \end{array}$$

but the particular choice of such an $\mathfrak{F}$ is of no immediate interest to us. As with all such terminological abuses this should cause no confusion.

For stable Postnikov systems the results of the previous section may be carried to a more complete stage. We shall again have to draw on results of Cartan to get us started.

Definition: Let A be an algebra over the Steenrod algebra $\mathcal{A}^*(p)$, p an odd prime. Define a morphism

$$\beta P_p : QA^- \longrightarrow QA^+$$

by

$$\beta P_p(x) = \beta P_p^t x \quad | \quad 2t+1 = \deg x .$$

<u>Proposition 5.1 (Cartan)</u>: Let A be a strictly positively graded abelian group. Then

$$\beta P_p : QH^*(KA; \mathbb{Z}_p)^- \longrightarrow QH^*(KA; \mathbb{Z}_p)^+$$

is a monomorphism.

<u>Proof</u>: This is a part of [10; Exposé 16; 2.1]. $\square$

<u>Notation</u>: As in the previous section p will denote a prime and $H^*(\ ; \mathbb{Z}_p)$ will be abbreviated to $H^*(\ )$.

From the previous section we now obtain:

<u>Theorem 5.2</u>: Let

$$\mathfrak{F} \qquad \begin{array}{ccc} E & \longrightarrow & L(A) \\ \downarrow & & \downarrow \\ X & \xrightarrow[f]{} & K(A) \end{array}$$

be a stable Postnikov system. Denote by $\{E_r(\mathfrak{F}), d_r(\mathfrak{F})\}$, the $\mathbb{Z}_p$-cohomology spectral sequence of $\mathfrak{F}$. Then

$$E_p(\mathfrak{F}) = E_\infty(\mathfrak{F}).$$

<u>Proof</u>: First note that $K(A)$ is 1-connected as A is a 1-connected graded abelian group. The finite type conditions are met and thus $\{E_r(\mathfrak{F}), d_r(\mathfrak{F})\}$ is defined.

The case $p = 2$ is contained in 3.2 so we may as well assume $p > 2$. According to 4.7 we have an isomorphism

$$E_2(\mathfrak{F}) = H^*(X)/\!/f^* \otimes E[s^{-1,0}QH^*(KA)\backslash\!\backslash f^{*+}] \otimes \Gamma[s^{-1,0}QH^*(KA)\backslash\!\backslash f^{*-}]$$

as Hopf algebras over $\mathcal{A}^*(p)$. The first non-zero differential is $d_{p-1}(\mathfrak{F})$ which is given by the formula

$$d_{p-1}(\mathfrak{F})(\gamma_j(s^{-1,0}x)) = (\lambda s^{-1,0}\beta P_p^t x)(\gamma_{j-1}(s^{-1,0}x))$$

where $j > 0$, $\lambda \neq 0 \in \mathbb{Z}_p$, and $x \in QH^*(KA)\backslash\!\backslash f^{*\pm}$.

We shall find the following computation of use.

Lemma: Consider the differential Hopf algebra

$$\mathcal{H} = \{ E[u] \otimes \Gamma[v] , d \}$$

where

$$d\, \gamma_j(v) = u\, \gamma_{j-1}(v) \qquad \text{for } j > 0$$

$$d\gamma_0(v) = 0 = du.$$

Then

$$H(\mathcal{H}) \cong P[v]/(v^p)$$

as a Hopf algebra over $\mathbb{Z}_p$.

Proof: Introduce the new variable

$$w = \gamma_p(v).$$

Note that as a differential algebra

$$\mathcal{H} \cong \{ E[u] \otimes \Gamma[w] \otimes \frac{P[v]}{(v^p)} , \quad d' \}$$

where

$$d'\, v = 0 = d'u$$

$$d'\, \gamma_i(w) = u\, \gamma_{i-1}(w) \qquad i \geq 0.$$

Thus we find

$$\mathcal{H} \cong \mathcal{E} \otimes \frac{P[v]}{(v^p)}$$

as differential algebras, where $\mathcal{E}$ is a Koszul complex.

Hence

$$H(\mathcal{H}) = P[v] / (v^p)$$

as algebras. As $P[v] / (v^p)$ has a unique Hopf algebra structure the result follows. $**$

Now by 5.1

$$\beta P_p : QH^*(K(A))^- \longrightarrow QH^*(K(A))^+$$

is monic . Choose a direct sum splitting

$$QH^*(KA)\backslash\backslash f^{*+} = \operatorname{Im} \beta P_p \oplus \operatorname{Coker} \beta P_p$$

as $\mathbb{Z}_p$-modules. We then find

$$E_{p-1}(\mathfrak{F}) = \cdots = E_2(\mathfrak{F}) \cong (H^*(X) /\!/ f^* \otimes E[s^{-1,0}\operatorname{Coker} \beta P_p])$$

$$\otimes (E[s^{-1,0}\operatorname{Im} \beta P_p] \otimes \Gamma[s^{-1,0}QH^*(K(A))\backslash\backslash f^{*-}]).$$

Thus we find that $E_{p-1}(\mathfrak{F})$ is the tensor product of a differential Hopf algebra of differential zero, and tensor products of differential Hopf algebras of the form of the lemma. Hence we find

$$E_p(\mathfrak{F}) \cong H^*(X) /\!/ f^* \otimes E[s^{-1,0}\operatorname{Coker} \beta P_p] \otimes \frac{P[s^{-1,0}QH^*(KA)\backslash\backslash f^{*-}]}{(s^{-1,0}QH^*(KA)\backslash\backslash f^{*-})^p}$$

as Hopf algebras over $\mathbb{Z}_p$.

Thus we find that as a $\mathbb{Z}_p$-algebra, $E_p(\mathfrak{F})$ is generated by

$$H^*(X)\backslash\backslash f^* = E_p^{0,*}(\mathfrak{F})$$

$$s^{-1,0}\operatorname{Coker} \beta P_p \subset E_p^{-1,*}(\mathfrak{F})$$

$$s^{-1,0}QH^*(KA)\backslash\backslash f^{*-} \subset E_p^{-1,*}(\mathfrak{F}) .$$

Now for any $r \geq p$ (> 2) we will have

$$d_r(\mathfrak{F}) : E_p^{0,*}(\mathfrak{F}) \longrightarrow E_r^{r,*}(\mathfrak{F}) = 0$$

$$d_r(\mathfrak{F}) : E_p^{-1,*}(\mathfrak{F}) \longrightarrow E_r^{r-1,*}(\mathfrak{F}) = 0 .$$

Thus for any $r \geq p$, $d_r(\mathfrak{F})$ must vanish on the algebra generators of $E_p(\mathfrak{F})$. As each $d_r(\mathfrak{F})$ is a derivation this implies $d_r(\mathfrak{F}) = 0$; $r \geq p$ and the result follows. □

Actually we have proved more than we have stated. We have actually computed $E_p(\mathfrak{F})$, and hence $E_\infty(\mathfrak{F})$, in the course of the proof. What we have shown is:

<u>Theorem 5.3</u>: Let

$$\mathfrak{F} \qquad \begin{array}{ccc} E & \longrightarrow & L(A) \\ \pi\downarrow & & \downarrow \\ X & \xrightarrow[f]{} & K(A) \end{array}$$

be a stable Postnikov system. Then there is a filtration $\{F^q H^*(E)\}$ such that

$$E_0 H^*(E) \cong H^*(X)/\!/f^* \otimes E[s^{-1,0}\mathrm{Coker}\,\beta P_p] \otimes \frac{P[s^{-1,0}QH^*(KA)\backslash\!\backslash f^{*-}]}{(s^{-1,0}QH^*(KA)\backslash\!\backslash f^{*-})^p}$$

as Hopf algebras over the Steenrod algebra $\mathcal{A}^*(p)$. □

Note the following instant Corollary.

<u>Corollary 5.4</u>: Let $\mathfrak{F}$ be as above. Then the natural map

$$H^*(X)/\!/ f^* \longrightarrow H^*(E)$$

is monic. □

It is upon Corollary 5.4 and a few extensions that we will build on generalization of the Cartan-Serre theorem. This will necessiate a more probing study of exactness and coexactness properties of Hopf algebras. This will be the subject of the next section. We will then have available all that we need for our generalization of the Cartan-Serre theorem, and will turn to it in full.

§6. Solvable and p-solvable Hopf Algebras

In the previous section we introduced the appropriate generalization of the Eilenberg-MacLane spaces that occur in the statement of the Cartan-Serre Theorem. We must now turn to the generalization of the coprimitive Hopf algebras that also occur in its formulation.

A more detailed discussion of these and related notions may be found in [31;I.2] .

Definition: Let k be a field. A homology Hopf algebra is a connected positively graded Hopf algebra over k whose comultiplication is commutative.

The category of homology Hopf algebras over k and their morphisms is denoted by $\mathcal{H}_*\mathcal{H}/k$.

Remarks on Gradings: When dealing with homology Hopf algebras we shall always use lower index notation in contrast to the upper index notation employed for cohomology Hopf algebras.

Remarks on Duality: If we were to demand that the underlying k-modules of our Hopf algebras must be of finite type then the categories $\mathcal{H}_*\mathcal{H}/k$ and $\mathcal{H}^*\mathcal{H}/k$ would be categorical duals. This duality would be consistent with our indexing conventions and the usual rules for raising and lowering indeces.

While the restriction to finite type will never be imposed, the heuristic duality that it suggests will often be invoked. Usually the dual proofs of assertions are valid without the restriction. The cautious reader may wish to make it part of the definitions. It will then be necessary to check that our constructions do not destroy it.

The dual of 2.1 becomes:

Proposition 6.1: Let A', A" ε obj $\mathcal{H}_*\mathcal{H}/k$ and $f : A' \longrightarrow A"$ a morphism of Hopf algebras. Then

(1) the natural morphism

$$j : k \,\square_{A"}\, A' \longrightarrow A'$$

is a kernel of f in $\mathcal{H}_*\mathcal{H}/k$, i.e., there is a natural homology Hopf algebra structure on $k \,\square_{A"}\, A'$ such that $j: k \,\square_{A"}\, A' \longrightarrow A'$ is a morphism of Hopf algebras satisfying the universal property of a kernel.

(2) If A" is abelian, i.e., the multiplication in A" is also commutative, then the natural morphism

$$c : A" \longrightarrow k \otimes_{A'} A"$$

is a cokernel of f in $\mathcal{H}_*\mathcal{H}/k$, i.e., there is a natural homology Hopf algebra structure on $k \otimes_{A'} A"$ such that $c : A" \longrightarrow k \otimes_{A'} A"$ is a morphism of Hopf algebras satisfying the universal property of a cokernel. □

We will continue to follow the notation for kernels and cokernels established in the second section.

Recollection: A homology Hopf algebra H is coprimitive iff $PH \longrightarrow QH$ is monic.

In this section we are seeking a generalization of the notion of coprimitive. Now recall (or compute) that if $x,y \in PH$ then $xy-(-1)^{\deg x \deg y}yx \in PH$ also. Thus certainly a coprimitive Hopf algebra must be abelian. This suggests that we introduce a measure of how non-abelian a homology Hopf algebra is. We turn to this now.

Definition: Let $H \in \operatorname{obj} \mathcal{H}_*\mathcal{H}/k$ and $J \subset H$ an ideal. Then we say that J is a Hopf ideal iff

$$\Delta(J) \subset H \otimes J + J \otimes H$$

where $\Delta : H \longrightarrow H \otimes H$ is the comultiplication in H.

Proposition 6.2: If $H \in \operatorname{obj} \mathcal{H}_*\mathcal{H}/k$ and $J \subset H$ is a Hopf ideal, then H/J has a natural homology Hopf algebra structure and

$$H \longrightarrow H/J$$

is a surjection of Hopf algebras.

Proof: This is elementary. □

Definition: Let $H \in \operatorname{obj} \mathcal{H}_*\mathcal{H}/k$. If $x, y \in H$ their commutator, $[x,y]$, is by definition

$$[x,y] = xy - (-1)^{\deg x \deg y}\, yx.$$

Definition: Let $H \in \operatorname{obj} \mathcal{H}_*\mathcal{H}/k$. The commutator ideal of H, denoted by $I_{[\,,\,]}$, is defined as the smallest ideal of H containing all the commutators of H.

Proposition 6.3: Let $H \in \operatorname{obj} \mathcal{H}_*\mathcal{H}/k$. Then the commutator ideal $I_{[\,,\,]} \subset H$ is a Hopf ideal.

Proof: An elementary computation. □

Definition: If $H \in \operatorname{obj} \mathcal{H}_*\mathcal{H}/k$, the abelianization of H, denoted by $C(H)$, is by definition $H/I_{[\,,\,]}$ together with the natural surjection of homology Hopf algebras

$$c(H) : H \longrightarrow H/I_{[\,,\,]} \equiv C(H) .$$

Proposition 6.4: Let $H \in \operatorname{obj} \mathcal{H}_*\mathcal{H}/k$. Then the morphism $c(H) : H \longrightarrow C(H)$ enjoys the following universal property.

Suppose that $A \in \text{obj}\,\mathcal{H}_*\mathcal{H}/k$ is an abelian Hopf algebra and $f : H \longrightarrow A$ is a morphism of Hopf algebras. Then there is a unique morphism

$$c(f) : C(H) \longrightarrow A$$

such that the diagram

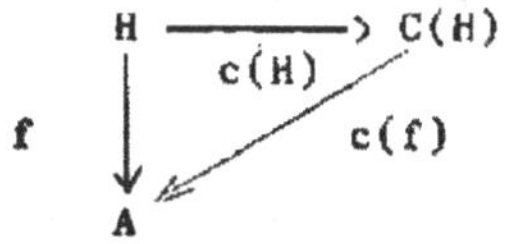

is commutative.

<u>Proof</u>: Immediate from the definitions, for since A is abelian f must vanish on the commutator ideal of H. The rest is routine. □

According to 6.1 every morphism in $\mathcal{H}_*\mathcal{H}/k$ has a kernal. Thus we may intreduce:

<u>Definition</u>: Let $H \in \text{obj}\,\mathcal{H}_*\mathcal{H}/k$. The commutator sub-Hopf algebra of H, denoted by either $[H,H]$ or $H[-1]$, is by definition the kernal of the natural map $c(H):H \longrightarrow C(H)$.

<u>Definition</u>: Let $H \in \text{obj}\,\mathcal{H}_*\mathcal{H}/k$. The derived series of H,

$$\ldots \subset H[-n-1] \subset H[-n] \subset \ldots \subset H[-1] \subset H \quad ,$$

is the descending series of homology Hopf algebras defined inductively by

$$H[-n-1] = (H[-n])[-1]$$
$$H[0] = H \quad .$$

<u>Definition</u>: A homology Hopf algebra H is solvable iff for some integer $n > 0$, $H[-n] = k$. If H is solvable

then the least integer n such that $H[-n] = k$ is called the index of solvability of H, and is denoted by $sol(H)$.

<u>Remark</u>: It is easy to extend the definition of C, $[-n]$, etc. to morphisms and thereby obtain functors

$$C : \mathcal{H}_*\mathcal{H}/k \longrightarrow \mathcal{H}_*\mathcal{H}/k$$
$$[-n] : \mathcal{H}_*\mathcal{H}/k \longrightarrow \mathcal{H}_*\mathcal{H}/k .$$

The key property of $sol(H)$ that will be needed for our generalization of the Cartan-Serre theorem is the sub-additivity of $sol(H)$ under extension. We will begin by recalling the basic notions of exactness and coexactness that we shall need.

<u>Recollections on Epimorphisms and Monomorphisms</u>: Let $\mathcal{A}$ be a category. A morphism in $\mathcal{A}$,

$$f : A' \longrightarrow A'',$$

is said to be an epimorphism if whenever

$$g', g'' : A'' \longrightarrow B$$

are morphisms in $\mathcal{A}$ such that $g'f = g''f$, then $g' = g''$. Dually, we say f is a monomorphism if whenever

$$h', h'' : C \longrightarrow A'$$

are morphisms in $\mathcal{A}$ such that $fh' = fh''$, then $h' = h''$.

These concepts are categorically dual.

<u>Proposition6.5.</u>: Let $f\ H' \longrightarrow H''$ be a morphism of homology Hopf algebras. Then

(1) f is an epimorphism iff the underlying morphism of k-modules induced by f is an epimorphism.

(2) f is a monomorphism iff the underlying morphism of k-modules induced by f is a monomorphism. □

<u>Proposition 6.5*</u>: Let $f : H' \longrightarrow H''$ be a morphism of cohomology Hopf algebras. Then

(1) f is an epimorphism iff the underlying morphism of k-modules induced by f is an epimorphism.

(2) f is a monomorphism iff the underlying morphism of k-modules induced by f is a monomorphism. □

<u>Remark</u>: It follows quite easily from 6.5_* that the functors

$$[-n] : \mathcal{H}_*\mathcal{H}/k \longrightarrow \mathcal{H}_*\mathcal{H}/k \qquad : n > 0 ,$$

preserve monomorphisms. They also preserve epimorphisms, although this is perhaps not so obvious. A proof of these facts may be found in [44].

<u>Warning</u>: These functors are not exact.

Thus for example, if H is a solvable homology Hopf algebra and $H' \subset H$ is a sub-Hopf algebra then H' is also solvable and

$$\text{sol}(H') \leq \text{sol}(H).$$

For if $i : H' \longrightarrow H$ is the inclusion, it is by 6.5_* a monomorphism. Hence

$$i[-n] : H'[-n] \longrightarrow H[-n]$$

is also monic, whence 6.5_* again yields the result.

<u>Recollections on Exact and CoExact Sequences:</u> Let $\mathcal{A}$ be a pointed category. A pair of morphisms

$$A' \xrightarrow{f'} A \xrightarrow{f''} A''$$

in $\mathcal{A}$ is called a two term sequence iff the composition $f''f' : A' \longrightarrow A''$ is trivial.

A two term sequence in $\mathcal{A}$,

$$A' \xrightarrow{f'} A \xrightarrow{f''} A'',$$

is called <u>exact</u> iff f'' has a kernal

$$\eta : \ker f'' \longrightarrow A$$

and the unique morphism

$$\zeta : A' \longrightarrow \ker f''$$

is an epimorphism.

Dually, a two term sequence

$$A' \xrightarrow{f'} A \xrightarrow{f''} A''$$

is called <u>coexact</u> iff f' has a cokernal

$$\xi : A \longrightarrow \operatorname{cok} f'$$

and the unique morphism

$$\lambda : \operatorname{cok} f' \longrightarrow A''$$

is a monomorphism.

Exactness and coexactness are categorically dual concepts, and the two are the same in an abelian category.

Exactness of sequences longer than two term are defined in terms of exactness of the two term components of the sequence.

If $* \in \operatorname{obj} \mathcal{A}$ is the point of $\mathcal{A}$ then the assertion

$$* \longrightarrow A' \xrightarrow{f'} A$$

is exact is equivalent to the condition that f' be a monomorphism, while

$$A \xrightarrow{f''} A'' \longrightarrow *$$

is coexact iff f'' is an epimorphism.

<u>Corollary 6.6.</u>: Let $f: H' \longrightarrow H''$ be a morphism of

homology Hopf algebras. If f is an epimorphism and a monomorphism then f is an isomorphism. □

<u>Corollary 6.6*</u>: Let $f : H' \longrightarrow H''$ be a morphism of cohomology Hopf algebras. If f is an epimorphism and a monomorphism then f is an isomorphism. □

<u>Corollary 6.7$_*$</u>: Let

$$k \longrightarrow H' \xrightarrow{f'} H \xrightarrow{f''} H'' \longrightarrow k$$

be an exact sequence of homology Hopf algebras. Then $f' = \ker f''$.

<u>Proof</u>: By definition we must have a commuatative diagram

$$\begin{array}{ccccc} k & \longrightarrow & H' & \xrightarrow{f'} & H \\ & & & \searrow^{g'} & \downarrow{\scriptstyle g''} \\ & & & & \ker f'' \end{array}$$

with g' an epimorphism. As f' is a monomorphism it follows from 6.5$_*$ that g' is also. Now apply 6.6$_*$. □

<u>Corollary 6.7*</u>: Let

$$k \longrightarrow H' \xrightarrow{f'} H \xrightarrow{f''} H'' \longrightarrow k$$

be a coexact sequence of cohomology Hopf algebras. Then $f'' = \operatorname{cok} f'$. □

<u>Proposition 6.8</u>: Suppose that

$$k \longrightarrow H' \xrightarrow{f'} H \xrightarrow{f''} H'' \longrightarrow k$$

is an exact sequence in $\mathcal{H}_*\mathcal{H}/k$, and H', H'' are solvable homology Hopf algebras. Then H is a solvable homology Hopf algebra and

$$\text{sol}(H) \leq \text{sol}(H') + \text{sol}(H'') .$$

<u>Proof</u>: Let $\text{sol}(H') = m'$, $\text{sol}(H'') = m''$. Then we have the diagram

$$
\begin{array}{ccccccccc}
 & & & & H[-m''] & \xrightarrow{f''[-m'']} & H''[m''] = k & & \\
 & & & & \downarrow{\scriptstyle i[m'']} & & \downarrow{\scriptstyle i''[-m'']} & & \\
k & \longrightarrow & H & \xrightarrow{f'} & H & \xrightarrow{f''} & H'' & \longrightarrow & k
\end{array}
$$

Thus the composite

$$f''i[-m''] : H[-m''] \longrightarrow H''$$

is the trivial morphism. By exactness

$$f' : H' \longrightarrow H$$

is the kernal of

$$f'' : H \longrightarrow H''$$

as follows from 6.7_* . Thus we receive a morphism

$$j : H[-m''] \longrightarrow H'$$

such that the diagram

$$
\begin{array}{ccccc}
 & & & & H[-m''] \\
 & & & \swarrow{\scriptstyle j} & \downarrow{\scriptstyle i[-m'']} \\
k & \longrightarrow & H' & \xrightarrow{f'} & H
\end{array}
$$

commutes. As f' is a monomorphism so is j by 6.5_* . Thus

$$j[-m] : H[-m''][-m'] \longrightarrow H'[-m'] = k$$

must also be a monomorphism. Hence by 6.5_* we must have

$$H[-m''-m'] = H[-m''][-m'] = k$$

completing the proof. □

Let us return now to to our considerations of how to generalize the notion of coprimitive homology Hopf algebra. Certainly there exist abelian Hopf algebras that are not coprimitive. The results of [30,§4] indicate that an additional factor to be taken into account is the Frobenous map. We recall this now.

<u>Definition</u>: Let $H \in \text{obj}\,\mathcal{H}_*\mathcal{H}/k$. Assume that the

characteristic of k is $p \neq 0$. Then for $n \in \mathbb{Z}$, $pn = 0\ (2)$ define

$$\xi : H_n \longrightarrow H_n \mid \xi(x) = x^p \ , \ x \in H_n \ .$$

<u>Proposition6.9</u>: Let k be a perfect field of characteristic $p \neq 0$. Then if $A \in \mathrm{obj}\,\mathcal{H}_*\mathcal{H}/k$ is an abelian Hopf algebra the k - submodule $\xi(A) \subset A$ is a sub homology Hopf algebra.

<u>Proof</u>: This is a routine consequence of the binomial theorem

$$(x + y)^p = x^p + y^p$$

mod p. The fact that $\xi(A)$ is a k-submodule of A is a consequence of the perfectness of k . □

<u>Definition</u> Let k be a perfect field of characteristic $p \neq 0$, and $H \in \mathrm{obj}\,\mathcal{H}_*\mathcal{H}/k$. The coprimitivization of H, denoted by $C^p(H)$, is by definition $C(H)//\xi(H)$, together with the natural surjection of homology Hopf algebras

$$c^p(H) : H \longrightarrow C^p(H) = C(H)//\xi(H) \ .$$

<u>Proposition 6.10</u>: Let k be a perfect field of characteristic $p \neq 0$ and $H \in \mathrm{obj}\,\mathcal{H}_*\mathcal{H}/k$. Then the morphism $c^p(H) : H \longrightarrow C^p(H)$ enjoys the following universal property.

Suppose that $C \in \mathrm{obj}\,\mathcal{H}_*\mathcal{H}//k$ is a coprimitive Hopf algebra and $f : H \longrightarrow C$ is a morphism of Hopf algebras. Then there is a unique morphism of Hopf algebras

$$c^p(f) : C^p(H) \longrightarrow C$$

such that the diagram

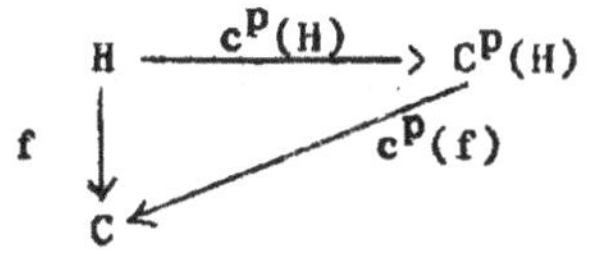

commutes. $\square$

According to 6.1 every morphism in $\mathcal{H}_*\mathcal{H}/k$ has a kernal. Thus we may introduce:

<u>Definition</u>: Let k be a perfect field of characteristac $p \neq 0$. Let $H \in \text{obj}\,\mathcal{H}_*\mathcal{H}/k$. We let $\langle H,H\rangle \subset H$ denote the kernal of the natural morphism $c^p(H) : H \longrightarrow C^p(H)$.

<u>Definition</u>: Let k be a perfect field of characteristic $p \neq 0$. Let $H \in \text{obj}\,\mathcal{H}_*\mathcal{H}/k$. The p-derived series of H,

$$\ldots \subset H\langle -n-1\rangle \subset H\langle -n\rangle \subset \ldots \subset H\langle -1\rangle \subset H$$

is the descending series of homology Hopf algebras defined inductively by

$$H\langle -1\rangle = \langle H, H\rangle$$

$$H\langle -n-1\rangle = \langle H\langle -n\rangle, H\langle -n\rangle\rangle .$$

<u>Definition</u>: A homology Hopf algebra H over a perfect field of characteristic $p \neq 0$ is p-solvable iff for some integer $n > 0$, $H\langle -n\rangle = k$. If H is p-solvable, then the least integer n such that $H\langle -n\rangle = k$ is called the index of p-solvability of H, and is denoted by p-sol(H).

Our final result in this section is the sub-additivity of p-sol( ). The proof is similar to 6.8 and is left to the reader.

<u>Proposition 6.11</u>: Let k be a perfect field of characteristic $p \neq 0$. Suppose that

$$k \longrightarrow H' \longrightarrow H \longrightarrow H'' \longrightarrow k$$

is an exact sequence in $\mathcal{H}/_*\mathcal{H}/k$, and H' , H" are p-solvable homology Hopf algebras. Then H is a p-solvable homology Hopf algebra and p-sol(H) $\leq$ p-sol(H') + p-sol(H") . □

7. The Generalized Cartan-Serre Theorem

We certainly now have at our disposal all that is needed to state our proposed generalization of the Cartan-Serre theorem. We make this our first order of bussisness.

Theorem 7.1: Let E be a Hopf space and p a prime integer. If E is a stable r-stage Postnikov system then $H_*(E ; \mathbb{Z}_p)$ is a p-solvable homology Hopf algebra and p-sol$(H_*(E ; \mathbb{Z}_p) \leq r$.

Theorem 7.2: Let E be a Hopf space and p a prime integer. If E is a stable r-stage Postnikov system then $H_*(E ; \mathbb{Z}_p)$ is a solvable homology Hopf algebra and sol$(H_*(E : \mathbb{Z}_p) \leq r$.

Remark: As noted in [31,II] there are Hopf spaces E with $H_*(E ; \mathbb{Z}_p)$ of arbitrarily high solvability and/or p-solvability, even though the underlying space of E is the homotopy type of K(A) for some A. The signifigance of the above above theorems in these cases is that in order to build up the Hopf space structure on such an E from the basic building blocks K(G,n) one needs several stages; at least as many as is called for by the solvability and p-solvability. Thus prop erly speaking 7.1, 7.2 is not a statement about Postnikov systems, but rather

Postnikov systems <u>in the category of Hopf spaces</u> or the Postnikov system of the classifying space.

As usual, there is a distinction between $p = 2$ and $p > 2$. As the case $p = 2$ is somewhat degenerate and the case $p>2$ more interesting we will consider only $p > 2$. The necessary changes in the details to accomodate the case $p = 2$ are left to the reader.

The following result is the key to the proof of 7.1 and 7.2.

<u>Proposition 7.3*</u> : Let

$$\mathfrak{F} \qquad \begin{array}{ccc} E & \longrightarrow & L(A) \\ \pi\big\downarrow & f & \big\downarrow \\ X & \longrightarrow & K(A) \end{array}$$

be a 1-connected Hopf fibre square of finite type. Let $\eta : K(s^{-1}A) \longrightarrow E$ be the inclusion of the fibre.

Define

$$R^* = H^*(X) /\!/ f^*$$

$$S^* = \operatorname{Im} \eta^* \subset H^*(K(s^{-1}A))$$

$$T^* = H^*(E) /\!/ \pi^* ,$$

where as usual $H^*(\ )$ denotes $H^*(\ ;\mathbb{Z}_p)$.

Then

$$\pi^* : R^* \longrightarrow H^*(E)$$

is a monomorphism, and there exists an exterior algebra on odd dimensional generators E^* such that

$$\mathbb{Z}_p \longrightarrow E^* \longrightarrow T^* \longrightarrow S^* \longrightarrow \mathbb{Z}_p$$

is coexact and T^* is primitive.

<u>Proof</u>: We will have to compare the $\mathbb{Z}_p$-cohomology

spectral sequence of $\mathcal{F}$ to that of

$$\mathcal{L} \qquad \begin{array}{ccc} K(s^{-1}A) & \longrightarrow & L(A) \\ \downarrow & & \downarrow \\ * & \longrightarrow & K(A) \end{array}$$

For this we shall need lemmas.

<u>Lemma</u>: Let $\theta : H^*(KA) \backslash\backslash f^* \longrightarrow H^*(KA)$ be the natural inclusion. Then

$$Q\,\theta^- : QH^*(KA)\backslash\backslash f^{*-} \longrightarrow QH^*(KA)^-$$

is monic.

<u>Proof</u>: By Cartan's theorem [10] $H^*(KA)$ is a free commutative algebra. By Borel's structure theorem [30;7.11] $H^*(KA)\backslash\backslash f^*$ must also be free commutative. The result then follows by inspection. $**$

<u>Lemma</u>: The natural map

$$\mathrm{Tor}_{H^*(KA)\backslash\backslash f^*}(\mathbb{Z}_p, \mathbb{Z}_p)^+ \longrightarrow \mathrm{Tor}_{H^*(KA)}(\mathbb{Z}_p, \mathbb{Z}_p)$$

induces a monomorphism

$$Q\mathrm{Tor}_{H^*(KA)\backslash\backslash f^*}(\mathbb{Z}_p, \mathbb{Z}_p)^+ \longrightarrow Q\mathrm{Tor}_{H^*(KA)}(\mathbb{Z}_p, \mathbb{Z}_p)^+$$

<u>Proof</u>: This follows from the previous lemma and the computation of the two torsion products by the Koszul complex construction, i.e., the discussion preceding 2.8. $**$

Let us consider now the natural morphism

$$\theta : \mathcal{L} \longrightarrow \mathcal{F}$$

of Hopf fibre squares defined by the diagram

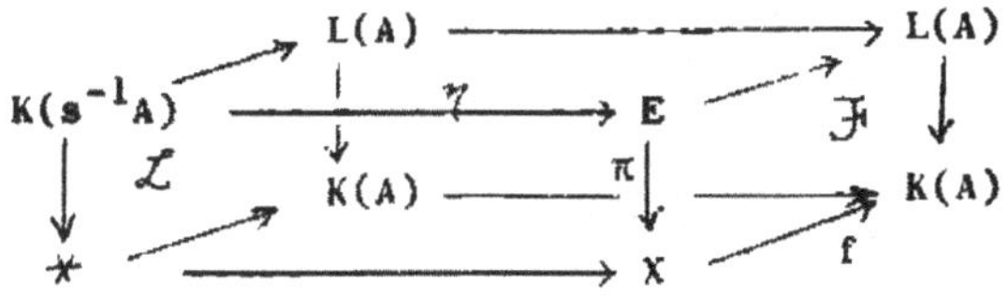

This induces a morphism of spectral sequences

$$\{E_r(\mathcal{F}),\ d_r(\mathcal{F})\} \longrightarrow \{E_r(\mathcal{L}),\ d_r(\mathcal{L})\}$$

that coincides with $\eta^* : H^*(E) \longrightarrow H^*(K\ s^{-1}A)$ on the limit and the natural map

$$R^* \otimes \operatorname{Tor}_{H^*(KA)\backslash\backslash f^*P}(\mathbb{Z}_p, \mathbb{Z}_p) \longrightarrow \operatorname{Tor}_{H^*(KA)}(\mathbb{Z}_p, \mathbb{Z}_p)$$

on the term E_2 .

Lemma: The natural morphism of homology Hopf algebras

$$E_\infty(\mathcal{F}) /\!/ R^* \longrightarrow E_\infty(\mathcal{L})$$

has kernel an exterior algebra on odd dimensional generators.

Proof: This is an easy consequence of the previous lemma and the structure of the E_∞ terms obtained in 5.3. **

Let us now examine our progress. Consider the map

$$\eta^* : H^*(E) \longrightarrow H^*(K(s^{-1}A)).$$

By 5.4

$$\pi^* : R^* \longrightarrow H^*(E)$$

is a monomorphism. Thus

$$H^*(E) /\!/ \pi^* = T^* = H^*(E) /\!/ R^*.$$

By the previous lemma, there is a filtration on $H^*(E) /\!/ R^*$ and $H^*(K(s^{-1}A))$ compatible with η^* and such that the induced morphism of homology Hopf algebras

$$E_0\, \eta^* : E_0H^*(E) /\!/ R^* \longrightarrow E_0H^*(K(s^{-1}A))$$

has kernel an exterior algebra on odd dimensional generators. Moreover

$$E_0^{0,*}\, H^*(E) /\!/ R^* = k$$

and

$$E_0^{-1,*}\, H^*(E) /\!/ R^* = PE_0H^*(E) /\!/ R^* .$$

Thus $H^*(E)//R^*$ is primitively generated. The proposition follows instantly. □

Now to prove 7.1 and 7.2 we shall need the homology version of 7.3*. Here is a statement of it. It follows by duality from 7.3*.

Proposition 7.3$_*$: Let

$$\mathcal{F} \quad \begin{array}{ccc} E & \longrightarrow & L(A) \\ \pi\downarrow & & \downarrow \\ B & \xrightarrow{f} & K(A) \end{array}$$

be a 1-connected Hopf fibre square of finite type. Let $\eta : K(s^{-1}A) \longrightarrow E$ be the inclusion of the fibre.

Define

$$R_* = H_*(X)\backslash\backslash f_*$$

$$S_* = \operatorname{Im} \eta_* \subset H_*(E)$$

$$T_* = H_*(E)\backslash\backslash \pi_* .$$

Then

$$\pi_*: H_*(E) \longrightarrow R_*$$

is an epimorphism, and there exists an exterior algebra E_* on odd dimensional generators such that

$$\mathbb{Z}_p \longrightarrow S_* \longrightarrow T_* \longrightarrow E_* \longrightarrow \mathbb{Z}_p$$

is exact and T_* is coprimitive. □

Proof of Theorems 7.1 and 7.2 : The proofs of these two results are similar and so we present them simultaneously. The case $r = 1$ is the theorem of Cartan [10] and thus we may assume $r > 1$. We proceed inductively and assume that the result has been established for $s < r$.

Choose a Hopf fibre square

$$\exists \quad \begin{array}{ccc} E & \longrightarrow & L(A) \\ {\scriptstyle\pi}\downarrow & & \downarrow \\ X & \xrightarrow{f} & K(A) \end{array}$$

that presents E as a stable r-stage Postnikov system. Then X is an r-1 stage Postnikov system. Thus by our inductive assumption $H_*(X)$ is solvable and p-solvable and

$$\text{sol } H_*(X) \leq r-1$$

$$\text{p-sol } H_*(X) \leq r-1.$$

By the remark following 6.5, it follows that R_* is solvable and p-solvable and

$$\text{sol } R_* \leq r-1$$

$$\text{p-sol } R_* \leq r-1.$$

Applying 6.8 and 6.11 to the exact sequence

$$\mathbb{Z}_p \longrightarrow T_* \longrightarrow H_*(E) \longrightarrow R_* \longrightarrow \mathbb{Z}_p$$

of 7.3, then yields

$$\text{sol } H_*(E) \leq r$$

$$\text{p-sol } H_*(E) \leq r$$

as T_* is coprimitive. This completes the inductive step and thus both results follow. □

This completes our discussion of the generalization of the theorem of Cartan and Serre discussed at the begining of the chapter.

§8. Closing Comments

In applying the Eilenberg-Moore spectral sequence to make the computations of this chapter we encountered several points we did not pursue and several open problems. We will try to describe some of these here.

Computations of $\mathrm{Tor}_\Lambda(B,A)$: As is already apparent from our work so far the first, and often difficult step, in using the Eilenberg-Moore spectral sequence in a given problem is the computation of the E_2-term. Namely, given morphisms of graded connected k-algebras

$$B \xleftarrow{\varphi} \Lambda \xrightarrow{\psi} A$$

to compute the bigraded k-algebra

$$\mathrm{Tor}_\Lambda(B,A).$$

Now as is already apparent, this is a non-trivial problem if B, Λ, A, φ and ψ possess no special additional structure. It is also apparent from our work that simply a knowledge of where the algebra generators of $\mathrm{Tor}_\Lambda(B,A)$ lie can be most useful in practice (e.g. 3.2).

Rather than leave this question in such an unstructured form, let me give an explicit problem, that grew out of discussions with P.F. Baum and J.C. Moore, and has an immediate bearing on geometric applications to the cohomology of homogeneous spaces of Lie groups [7].

We suppose that $B = k$ and A, Λ are polynomial algebras over k on even dimensional generators. No Hopf algebra structure is assumed or implied. It is also assumed that A and Λ are finitely generated as algebras.

(1) Is

$$\mathrm{QTor}_{\Lambda}(\,k,\ A)^{-s,*} = 0 \quad ;\ s > 2\ ?$$

(2) Is

$$[\ \mathrm{QTor}_{\Lambda}(\,k,\ A)^{-2,*}\,]^{-} = 0 \quad ?$$

<u>The Differential d_{p-1} and Adem Relations :</u> There is an interesting formula between Steenrod operations that follows from 4.7. Suppose that B is a 1-connected Hopf space. We then have the Hopf fibre square

$$\mathcal{P}(B) \qquad \begin{array}{ccc} \Omega B & \longrightarrow & PB \\ \downarrow & & \downarrow e \\ * & \longrightarrow & B \end{array}$$

where $e: PB \longrightarrow B$ is the path space fibration. Let p be an <u>odd</u> prime. Consider the $\mathbb{Z}_p$-cohomology spectral sequence $\{E_r(\mathcal{P}(B)),\ d_r(\mathcal{P}(B))\}$. Then

$$E_2(\mathcal{P}(B)) = \mathrm{Tor}_{H^*(B)}(\mathbb{Z}_p,\mathbb{Z}_p)$$

where as usual we have abbreviated $H^*(\ \ ;\mathbb{Z}_p)$ to $H^*(\ \ )$. Now using the Borel structure theorem and some more extensive computations than those of 2 we may show (see e.g.[31; II]):

$$E_2(\mathcal{P}(B))$$

$$= E[s^{-1,0}QH^*(B)^+] \otimes \Gamma[s^{-1,0}QH^*(B)^-] \otimes \Gamma[\bigoplus_{r=0}^{\infty} s^{-2,p^r} Q^{(r)}H^*(B)^+]$$

where $Q^{(r)}H^*(B)^+$ is defined by

$$x \in Q^{(r)}H^*(B)^+ \iff \begin{cases} x \in QH^*(B)^+ \text{ and} \\ x^{p^r} = 0 \in H^*(B) \text{ and } x^{p^{r-1}} \neq 0. \end{cases}$$

The procedure of §4 may be followed once again to deduce that $d_{p-1}(\mathcal{P}(B))$ is the first non-zero differential, and it is given by the formula

$$d_{p-1}(\mathcal{P}(B))(\gamma_j(s^{-1,0}x)) = \lambda s^{-1,0} \; P_p^t x \; \gamma_{j-1}(s^{-1,0}x)$$

for all $j \geq p$, where $x \in QH^*(B)^-$, $2t+1 = \deg x$ and $\lambda \neq 0 \in \mathbb{Z}_p$.

Now [46] the Steenrod algebra acts on E_2 and one may derive the formula

$$\gamma_p(s^{-1,0}P_p^s x) = P_p^{ps} \gamma_p(s^{-1,0} x)$$

for the action of the Steenrod algebra on $\mathrm{Tor}_{H^*(B)}(\mathbb{Z}_p, \mathbb{Z}_p) = E_2(\mathcal{P}(B))$. Applying $d_{p-1}(\mathcal{P}(B))$ yields

$$d_{p-1}(\mathcal{P}(B))(\gamma_p(s^{-1,0}P_p^s x))$$

$$= d_{p-1}(\mathcal{P}(B))(P_p^{ps} \gamma_p(s^{-1,0}x)).$$

Applying [42] to deduce the compatibility of the $\mathcal{A}^*(p)$ action with the differentials we obtain

$$d_{p-1}(\mathcal{P}(B))(P_p^{ps} \gamma_p(s^{-1,0}P_p^s x))$$

$$= P_p^{ps} d_{p-1}(\mathcal{P}(B))(\gamma_p(s^{-1,0} x)).$$

Applying the above formula for $d_{p-1}(\mathcal{P}(B))$ then yields the equation

$$\beta P^{t+s(p-1)} P_p^s x = P_p^{sp} \beta P_p^t x$$

in $QH^*(B)$. The case $t = s$ is

$$\beta P_p^{tp} P_p^t x = P_p^{tp} \beta P_p^t x$$

which follows from the Adam relations. I am not sure about the cases $t \neq s$.

At any rate, the formula 4.7 together with [42] can be seen to yield Adam type relations for the action of $\mathcal{A}^*(p)$ on $QH^*(B)$.

This point is important in the further study of the differentials $d_{p^r-1}(\mathcal{P}(B))$ for $r > 1$. We hope to give a complete discussion of these phenomena on another occasion.

Finally we note that in [40] a formula analogous to 4.7 is developed for $d_{2p-1}(\mathcal{P}(B))$.

Postnikov Systems: An r-stage Postnikov system is defined inductively by

(1) a 1-stage Postnikov system is a $K(A)$.

(2) For $r > 1$ an r-stage Postnikov system is a pair E, $\mathfrak{F}$, where E is a space and $\mathfrak{F}$ is a fibre square

$$\begin{array}{ccc} E & \longrightarrow & L(A) \\ {\scriptstyle\pi}\downarrow & & \downarrow \\ X & \xrightarrow{\ f\ } & K(A) \end{array}$$

where X is an $r-1$ stage Postnikov system.

The general problem is to compute the $\mathbb{Z}_p$-cohomology of such Postnikov systems. The case $r = 2$ is already sufficiently difficult to warrant a few remarks.

First note that 2-stage Postnikov systems serve as universal examples for secondary operations, stable operations corresponding to stable 2-stage Postnikov systems. Next note that it is the interplay between primary and secondary operations that is often most important in applications [3]. Thus what we would like to know is the action of the Steenrod algebra on the $\mathbb{Z}_p$-cohomology of a two stage Postnikov system E. As far as the $\mathbb{Z}_p$-algebra structure goes this may be drawn out of the results of § 5. This was done in [38] where complete details a may be found. Except in a few cases the action of $\mathcal{A}^*(p)$ remains a mystery. This seems like a difficult but important problem. The format of Massey and Peterson [27] seems well adapted to the problem.

There is also a question about what relations are forced by the choice of k-invariant f. Here is a specific such question.

Suppose that

$$\begin{array}{ccc} E & \longrightarrow & L(A) \\ \pi\downarrow & & \downarrow \\ K(B) & \xrightarrow[f]{} & K(A) \end{array}$$

is a 2-stage Postnikov system, not necessarily stable. For simplicity assume that A and B are graded $\mathbb{Z}_p$-modules, p a prime.

Let

$$H^*(K(B))/\!/f^* = \mathbb{Z}_p \otimes_{H^*(KA)} H^*(KB).$$

Is it true that

$$\pi^* : H^*(KB)/\!/f^* \longrightarrow H^*(E)$$

is monic? This amounts to asking if the ideals

$$\ker \pi^* \subset H^*(KB)$$

$$I^*(f) \equiv f^*\bar{H}^*(KA)\cdot H^*(KB) \subset H^*(KB)$$

are equal. Note $I^*(f) \subset \ker \pi^*$.

In case E is stable this is just 5.4 (see also [38]). Dr. Claude Schochet of the University of Chicago has obtained such results in the nonstable case (Thesis 1969).

CHAPTER III

The Cobar Construction: Applications

The first spectral sequence of Eilenberg-Moore type was discovered by J.F.Adams [1]. As originally developed [26] and [12] it applied to the pathspace fibration over a 1-connected space. The construction was algebraic in nature and introduced for the first time the cobar construction. It is our purpose to review in this chapter the original work of Adams and exploit various properties of the cobar construction to deduce two general results. This utilization of functorial properties of an algebraic construction of an Eilenberg-Moore type spectral sequence should be contrasted with our exploitation of the geometric construction of I and [43;I] in [43;II] and [46]. Both the algebra and the geometry have advantage to offer and neither should be ignored.

The first application of the cobar spectral sequence that we will present deals with the following situation. We suppose given a cofibration

$$A \hookrightarrow X \longrightarrow C$$

and we ask for the relation between

$$H_*(\Omega A; k) \quad H_*(\Omega X; k) \quad H_*(\Omega C; k)$$

where k is a field. We will find that when A, X and C are 1-connected there is a first quadrant homology spectral sequence

$$E^r \Longrightarrow H_*(\Omega X; k)$$

$$E^2 = H_*(\Omega A; k) \amalg H_*(\Omega C; k)$$

where $\amalg$ denotes the coproduct of the homology Hopf algebras $H_*(\Omega A;k)$ and $H_*(\Omega C; k)$. In the special case

$$X = A \vee C$$

we will find the spectral sequence collapses with trivial extension, reobtaining a result of [14].

The second situation we consider revolves around another theorem of Cartan and Serre [30; Appendix]. What Cartan-Serre showed is that for any H-space X the natural mapping

$$\pi_*(X) \otimes_{\mathbb{Z}} Q \longrightarrow PH_*(X; Q)$$

is an isomorphism, where Q is the rational numbers. The question we are concerned with is how does this result generalize to arbitrary spaces, not just H-spaces. What we shall show is that for 1-connected X the cobar spectral sequence readily yields up a spectral sequence

$$E^r \Longrightarrow \pi_*(X) \otimes_{\mathbb{Z}} Q$$

$$E^2_{-p,*} = R^{-p-1}\, PH_*(X; Q)$$

where $R^s P$ — is the sth right relative derived functor of P. The necessary algebraic preliminaries will also allow us to reinterpret the spectral sequence of a cofibration mentioned above in term of rational homotopy.

Most of the results of this chapter are an outgrowth of (unpublished) joint work with Alan Clark and it is a lpeasure to acknowledge the many useful and informative conversations that we have had. Most of these results are also known to I. Berstein, J.C.Moore and presumably others. However to the best of our knowledge this is the first time these results have found their way into print. We apologize for not giving as much detail in this chapter as in the previous ones. There is nothing significant left out, only tedium.

§ 1. The Cobar Spectral Sequence

Let us begin by reviewing the cobar construction as introduced by Adams [1]. As in previous chapters k denotes a fixed field.

Notation and Conventions: For any k-module M and integer p we denote by $s^p M$ the bigraded k-module given by

$$(s^p M)_{r,s} = \begin{cases} 0 & r \neq p \\ M_{s-p} & r = p \end{cases} .$$

The special case $p = -1$ will occupy us somewhat in the sequel and we introduce the notation [x] for the element of $s^{-1} M$ corresponding to $x \in M$.

Definition: Let C be a differential coalgebra over the field k. The cobar construction of C, denoted by $\mathfrak{F}(C)$ is the second quadrant differential Hopf algebra defined as follows:

(1) As an algebra $\mathfrak{F}(C)$ is the bigraded tensor algebra (i.e., free associative algebra)

generated by $s^{-1}JC$ where

$JC = \text{coker}\{ k \xrightarrow{\eta} C \}$; $\eta : k \longrightarrow C$ the counit of C;

(2) the diagonal of $\overline{\mathfrak{F}}(C)$ is determined uniquely by the requirement that

$$s^{-1}JC \subset P\overline{\mathfrak{F}}(C) \quad ;$$

(3) we define differentials d_I and d_E of $\overline{\mathfrak{F}}(C)$ of degrees $(0,-1)$ and $(-1,0)$ respectively, by setting

$d_I([c]) = -[dc]$: where $d : C \longrightarrow C$ is the differential of C and $c \in C$,

$d_E([c]) = \sum (-1)^{\deg c_i'} [c_i'|c_i'']$: where $\Delta : C \longrightarrow C \otimes C$ is the co-multiplication and $\Delta(c) = \sum c_i' \times c_i''$

and requiring that d_I and d_E be derivations of the algebra structure.

<u>Notation</u>: Following long established tradition we have denoted the element

$$[c_1][c_2] \ldots [c_n]$$

of $\overline{\mathfrak{F}}(C)$ by

$$[c_1|c_2|\ldots\ldots|c_n] \quad .$$

<u>Remarks</u>: (1) $d_I^2 = 0$ for elementary reasons;

(2) $d_E^2 = 0$ is a consequence of the associativity of Δ ;

(3) $d_I d_E + d_E d_I = 0$ because of the formula $d\Delta = \Delta(d \otimes 1 + 1 \otimes d)$

(4) the total differential on $\overline{\mathfrak{F}}(C)$ is defined by $d_T = d_I + d_E$.

The total cobar construction of C, denoted by $\mathcal{F}(C)$, is defined as follows. As a k-coalgenra

$$\mathcal{F}(C) = C \otimes_k \overline{\mathcal{F}}(C) .$$

We define differentials d_i and d_ε for $\mathcal{F}(C)$ by requiring that the diagrams

$$\begin{array}{ccc} C & \xrightarrow{\tau} & C \otimes \overline{\mathcal{F}}(C) \\ {\scriptstyle d}\downarrow & & \downarrow {\scriptstyle d_i} \\ C & \xrightarrow{\tau} & C \otimes \overline{\mathcal{F}}(C) \end{array}$$

$$\begin{array}{ccc} C & \xrightarrow{\tau} & C \otimes \overline{\mathcal{F}}(C) \\ \downarrow & & \downarrow {\scriptstyle d_\varepsilon} \\ C & \xrightarrow{\tau} & C \otimes \overline{\mathcal{F}}(C) \end{array}$$

commute, where τ is the natural k-morphism of degree -1 given by $C \xrightarrow{\cong} s^{-1}C \subset (C)$.

The total differential of $\mathcal{F}(C)$ is given by $d_\tau = d_i + d_\varepsilon$.

Proposition 1.1: Let C be a differential k-coalgebra. Then the differential k-coalgebra $(\mathcal{F}(C), d_\tau)$ is acyclic.

Proof: There are numerous ways to establish this result. One such is to exhibit an explicit contracting homotopy for $(\mathcal{F}(C), d_\tau)$. We leave to the reader to verify that the map

$$S : \mathcal{F}(C) \longrightarrow \mathcal{F}(C)$$

defined by

$$S(c[c_1|\ldots|c_n] = \varepsilon(c)c_1[c_2|\ldots|c_n]$$

is such a contracting homotopy, where $\varepsilon : C \longrightarrow k$ is the augmentation of C. □

It is readily verified that the pair

$$\tau : C \longrightarrow \overline{\mathcal{F}}(C) \quad , \overline{\mathcal{F}}(C)$$

allows us to view $(\mathcal{F}(C), d_\tau)$ as a twisted tensor product in the sense of Brown [9]. In view of the main result of [9] the following theorem of Adams is not suprising.

Notation: Suppose that X is a 1-connected space with a non-degenerate basepoint $x_0 \varepsilon X$. Let $C_*(X,x_0;k)$ or simply $C_*(X)$ denote the singular chain complex of X with all one simplices at the basepoint and coefficents in the field k.

Theorem 1.2(J.F. Adams) : Let X be a 1-connected space and $C = C_*(X)$. Then the spectral sequence obtained by filtering $(\mathcal{F}(C), d_\tau)$ by the first degree coincides with the Serre spectral sequence of the path space fibration

$$\Omega X \longrightarrow PX \longrightarrow X .$$

Hence there is a natural isomorphism of Hopf algebras

$$H_*(\Omega X) = H(\overline{\mathcal{F}}(C), d_T) . \square$$

As the datails of proof of 1.2 are quite similar to those of [18] we refer the reader [18] or [1], or suggest the reader construct a proof similar to theone we hinted at based on [9].

The next few sections are devoted to exploiting several functoriality properties of the cobar construction to deduce various geometric consequences on the basis of 1.2. For example the main result of [14] depends simply on the fact that $\overline{\mathcal{F}}(\ )$ preserves coproducts.

2 Cofibrations and $\overline{\mathcal{F}}(C)$

Before stating the main result of this section let us recall some elementary facts.

<u>CoProducts</u>: Let $\mathcal{C}$ be a category and A , $B \in \text{obj}\,\mathcal{C}$. A coproduct of A and B in $\mathcal{C}$ consists of a diagram in $\mathcal{C}$

$$A \xrightarrow[i_A]{} A \amalg B \xleftarrow[i_B]{} B$$

such that for any pair of morphisms

$$f : A \longrightarrow C \longleftarrow B : g$$

in $\mathcal{C}$ there exists a unique morphism

$$f \amalg g : A \amalg B \longrightarrow C$$

such that the diagram in $\mathcal{C}$

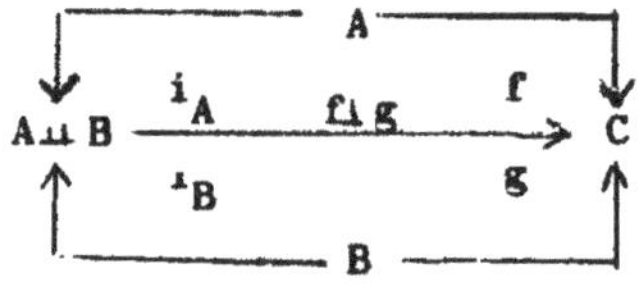

is commutative. As usual with universal gadgets if a coproduct of A and B exists in $\mathcal{C}$ it is unique up to a canonical isomorphism.

A category $\mathcal{C}$ is said to have coproducts if each pair of objects in $\mathcal{C}$ has a coproduct.

The following list of categories and their coproducts will be useful in the sequel.

<u>The Category $A\ell^+/k$ of Connected Positively Graded k-Algebras:</u> For $A \in \text{obj}A\ell^+/k$ we define a graded k-module $\bar{A}$ by

$$(\bar{A})_n = \begin{cases} A_n & \text{if } n > 0 \\ 0 & \text{if } n = 0 . \end{cases}$$

Then $T(\bar{A})$, the tensor algebra of $\bar{A}$ is an object of $A\ell^+/k$, and there is a canonical morphism

$$T(\bar{A}) \longrightarrow A$$

in $A\ell^+/k$ whose kernal we denote by IA (kernal in the sense of morphisms of k-modules!).

If A , $B \in \text{obj}\, A\ell^+/k$, their coproduct is defined by

$$A \amalg B = T(\bar{A} \oplus \bar{B})/(IA\ ,\ IB)$$

(where the denominator denotes the smallest ideal of $T(\bar{A} \oplus \bar{B})$ containing IA and IB) together with the obvious maps

$$A \longrightarrow T(\bar{A} \oplus \bar{B})/(IA, IB) \longleftarrow B\ .$$

The universal property of the coproduct is easily verified.

<u>The Category $\mathcal{H}^+/k$ of Connected Positively Graded Hopf Algebras over k</u>: If A , $B \in \text{obj}\,\mathcal{H}^+/k$ then we may regard them as objects of $A\ell^+/k$ and form their coproduct $A \amalg B$ in $A\ell^+/k$. In $A\ell^+/k$ we may construct the diagram

$$\begin{array}{ccccc}
A & \xrightarrow{\quad i_A \quad} & A \amalg B & \xleftarrow{\quad i_B \quad} & B \\
\downarrow{\scriptstyle \Delta_A} & & \downarrow{\scriptstyle \Delta_{A \amalg B}} & & {\scriptstyle \Delta_B}\downarrow \\
A \otimes A & \xrightarrow{\quad i_A \times i_A \quad} & (A \amalg B) \otimes (A \amalg B) & \longleftarrow & B \otimes B
\end{array}$$

defining the morphism $\Delta_{A \amalg B}$ by the universal property of the coproduct in $A\ell^+/k$ and imposing a Hopf algebra structure on $A \amalg B$. Equipped with this coproduct one readily checks that $A \amalg B$ satisfies the universal property of the coproduct in $\mathcal{H}^+/k$.

<u>The Category $\mathcal{H}_*\mathcal{H}/k$:</u> One may readily verify that the natural forgetful functor

$$\text{For}: \mathcal{H}^+/k \longrightarrow \mathcal{H}_*\mathcal{H}/k$$

takes coproducts to coproducts. Thus the coproduct in $\mathcal{H}_*\mathcal{H}/k$ may be defined by the same procedure as in $\mathcal{H}^+/k$.

<u>The Category $D\mathcal{H}^+/k$ of Differential Connected Positively</u>

Graded Hopf Algebras

The coproduct is defined as above, the differential extends naturally.

We will now state the main result of this section.

Theorem 2.1: Suppose that

$$A \hookrightarrow X \longrightarrow B$$

is a cofibration sequence of 1-connected spaces, and k is a field. Then there exists a natural spectral sequence of Hopf algebras over k.

$$E^r \Longrightarrow H_*(\Omega X;\, k)$$

$$E^2_{p,q} \cong [\, H_*(\Omega A;\, k) \amalg H_*(\Omega B;\, k\,)]_{p,q}$$

where $\amalg$ denotes the coproduct in the category $\mathcal{H}_*\mathcal{H}/k$.

Proof: Let us begin with the exact sequence

$$C_*(A) \xrightarrow{i_*} C_*(X) \xrightarrow{p_*} C_*(B)$$

of differential k coalgebras. Applying the functor $\mathfrak{F}$ leads to the sequence of differential k-Hopf algebras

$$\mathfrak{F}C_*(A) \xrightarrow{\mathfrak{F}i_*} \mathfrak{F}C_*\ (X) \xrightarrow{\mathfrak{F}p_*} \mathfrak{F}C_*(B)\ .$$

Let us define a filtration on $\mathfrak{F}C_*(X)$ as follows. First define a filtration on $\mathfrak{F}C_*(B)$ by the cobar degree, i.e.,

$$x \in F_n\mathfrak{F}C_*(B) \iff x = \sum_{m \geq n} x_i \mid x_i = [c_{1,i}|\cdots\cdots|c_{m,i}]$$

and setting

$$F_n\mathfrak{F}C_*(X) = \mathfrak{F}p_*^{-1}(F_n\,\mathfrak{F}\,C_*(B)).$$

It is our contention that the spectral sequence of this

filtration has the properties asserted in the theorem. The verification is based on several steps, depending ultimately on the structure of $E^0 \mathfrak{F} C_*(X)$.

So our first step is to examine $E^0 \mathfrak{F} C_*(X)$. To this end choose a map of k-modules

$$\sigma : C_*(B) \longrightarrow C_*(X)$$

splitting the k-module morphism p_*. This is possible because p_* is onto and $C_*(B)$ is a free k-module. Note that σ need not be a differential morphism.

For each element $x \in C_*(X)$ we denote by

$$[x] \in E^0_{1,*} \mathfrak{F} C_*(X)$$

the corresponding element of $\mathfrak{F} C_*(X)$. The k-module morphisms 1_* and σ then combine to yield morphisms of k-modules

$$C_*(A) \longrightarrow E^0 \overline{} C_*(X) \longleftarrow C_*(B).$$

From the fundamental properties of the tensor or free algebra construction we thus obtain morphisms of k-algebras

$$\mathfrak{F} C_*(A) \longrightarrow E^0 \mathfrak{F} C_*(X) \longleftarrow \mathfrak{F} C_*(B)$$

which via the universal property of $\amalg$ yields a morphism

$$\varphi : \mathfrak{F} C_*(A) \amalg \mathfrak{F} C_*(B) \longrightarrow E^0 \mathfrak{F} C_*(X)$$

of Hopf algebras over k.

Now we will regard $\mathfrak{F} C_*(A) \amalg \mathfrak{F} C_*(B)$ as bigraded by using degrees of $\mathfrak{F} C_*(B)$ as filtration degree and those of $\mathfrak{F} C_*(A)$ as fibre degree. We impose a differential on $\mathfrak{F} C_*(A) \amalg \mathfrak{F} C_*(B)$ by extending the trivial derivation on $\mathfrak{F} C_*(B)$ and the total derivation on $\mathfrak{F} C_*(A)$. It is then verified that the homomorphism φ is a morphism of differential Hopf algebras over k. Moreover, as φ is dearly an isomorphism of k-algebras without differential we have obtained the following fundamental fact

$$E^0 \mathfrak{F}C_*(X) \cong (\mathfrak{F}C_*(A) \amalg \mathfrak{F}C_*(B), \quad d_\tau \amalg "0")$$

as differential Hopf algebras over k.

It follows from the explicit construction of $\mathfrak{F}C_*(A) \amalg \mathfrak{F}C_*(B)$ and the classical Künneth theorem over k that

$$E^1 \mathfrak{F}C_*(X) \cong H_*(\mathfrak{F}C_*(A)) \amalg \mathfrak{F}C_*(B)$$

The differential d^1 is easily seen to be the extension of the trivial derivation on $H_* \mathfrak{F}C_*(A)$ and the total differential on $\mathfrak{F}C_*(B)$. Thus appealing to the Künneth theorem again we find that

$$E^2 \mathfrak{F}C_*(X) \cong H_*(\mathfrak{F}C_*(A)) \amalg H_*(\mathfrak{F}C_*(B)).$$

As the filtration is complete [17] [18] we find that

$$E^r \Longrightarrow H(\mathfrak{F}C_*(X))$$

in the strong sense. Appealing to the theorem of Adams [1] (Theorem 1.2 above) to reidentify $H(\mathfrak{F}C_*(Y))$ with $H_*(\Omega Y; k)$ for 1-connected Y complete the proof.□

As an interesting consequence we note the following:

Corollary 2.2: Suppose that the cofibration

$$A \xrightarrow{i} X \xrightarrow{p} B$$

admits a cocross-section

$$s : B \longrightarrow X .$$

Then for any field k the natural map

$$s_* \perp i_* : H_*(\Omega B; k) \amalg H_*(\Omega A; k) \longrightarrow H_*(\Omega X;k)$$

is an isomorphism.

Proof: The usual spectral sequence argument that shows the spectral sequence of 2.1 collapses with trivial extension. The situation is similar to that of the Serre spectral sequence of a fibration with cross-section. □

The special case of $X = A \vee B$ was examined in some detail as a step towards the main result of [14].

From a theoretical standpoint the spectral sequence of 2.2 is quite pretty. However it is of quite non-standard form in that the term E_2 is not anything like the tensor product of the base and fibre terms. For this reason none of the usual spectral arguments involving exact sequences of terms of low degree, two term conditions, etc. seem to have any meaningful analogs. One would seem to require a whole new bag of tricks to make a living with this spectral sequence.

We close this section with an elementary application of 2.2 .

Recollection: Let k be a field and M a graded k-module of finite type. The Poincaré series of M is the formal power series defined by

$$P(M,t) = \sum \dim_k(M_n)t^n \quad .$$

Definition: Let k be a field. For any 1-connected space of finite type define

$$w(X) = P(H_*(\Omega X;k) \; , \; t) \; ,$$

the Poincaré series of the mod k homology of the loop space of X . Let

$$w(X) = 1 + \tilde{w}(X)$$

<u>Proposition 2.3</u>: For any pair of pointed 1-connected spaces A and B of finite type we have

$$\widetilde{w}(A \vee B) = \frac{\widetilde{w}(A) + \widetilde{w}(B)}{1 - \widetilde{w}(A)\widetilde{w}(B)} .$$

<u>Proof</u>: By direct computation from the explicit construction of the coprodact of homology Hopf algebras and 2.2 . □

§ 3. On Differential Hopf Algebras Over Q.

Our objective in this section is to examine some special properties of differential Hopf algebras over the rational number field Q. These will be applied in the next, and final section to deduce various results on rational homotopy.

Let (A,d) be a differential Hopf algebra over the field k. It is immediate that

$$d(PA) \subset PA$$

where as usual $PA \subset A$ denotes the submodule of primitive elements of A. We thus obtain a differential Lie algebra over k, $(PA, d|_{PA})$. Hence there is a natural map of Lie algebras over k

$$\rho_A : HPA \longrightarrow PHA$$

induced by the inclusion of complexes over k

$$(PA, d|_{PA}) \hookrightarrow (A,d).$$

It is our intension to study the problem of when

$$\rho_A : HPA \longrightarrow PHA$$

is an isomorphism. The main result that we shall establish is :

<u>Theorem 3.1</u> : Let (A,d) be a differential homology

Hopf algebra over the rational number field Q. Then

$$\rho_A : HPA \longrightarrow PHA$$

is an isomorphism

By a differential homology Hopf algebra we of course understand a differential Hopf algebra whose underlying Hopf algebra is homology Hopf algebra.

Actually it will be more convenient to establish the result dual to 3.1, dealing with the natural mapping

$$q_A : QHA \longrightarrow HQA$$

defined for any differential Hopf algebra (A,d). The details for 3.1, are left to the coreader.

<u>A Filtration</u>: We assume that (A,d) is a differential Hopf algebra over the field k and introduce the filtration

$$\cdots\; F^p A \subset \cdots\cdots \subset F^1 A \subset F^0 A = A$$

by setting

$$F^0 A = A$$

$$F^p A = IA \cdot F^{p-1} A \qquad p > 0 \text{ and } IA \hookrightarrow A \text{ the augmentation ideal}$$

One readily checks that

$$d\, F^p A \subset F^p A$$

and hence (A,d) becomes a filtered differential k-module, filtered by

$$\cdots\; (F^p A, d|_{F^p A}) \subset \cdots \subset (F^1 A, d|_{F^1 A}) \subset (A,d).$$

There results a spectral sequence $\{E_r(A), d_r(A)\}$ where

$$E_r(A) \Longrightarrow H(A,d)$$

and

$$E_1(A) = H(E_0A, E_0d).$$

We note that

$$E_0^{0,*}(A) \cong k$$

$$E_0^{1,*}(A) \cong QA$$

and thus we have induced an edge type map

$$HA \longrightarrow E_\infty^{1,*} \longrightarrow E_1^{1,*} = H(QA)$$

which factors thru

$$HA \longrightarrow QHA$$

such that the induced map

$$QHA \longrightarrow HQA$$

coincides with the natural mapping q_A.

Recollections: (1) The filtration $\{F^pA\}$ was originally introduced by Milnor-Moore. It is readily checked that [30]

$$Q\,E_0 = E_0^{1,*}A \cong s^{-1}\,QA$$

and that

$$PE_0A \longrightarrow QE_0A$$

is epic, i.e., E_0A is a primitive Hopf algebra. Moreover, if A is coprimitive one finds that $E_0\,A$ is also coprimitive. Hence if A is coprimitive then $E_0\,A$ is biprimitive.

(2) Over the rational number field Q the coprimitive Hopf algebras are exactly the commutative ones [30]. Moreover, a coprimitive Hopf algebra over Q is isomorphic as an algebra to a free commutative algebra.

(3) Let M be a graded module over the field k. We define the Poincaré series to be the formal power series given

by

$$P(M,t) = \sum_{i=0}^{\infty} \dim_k(M_i)\, t^i \quad .$$

We recall that for free commutative algebras over k, A,B that $A \cong B$ as graded k-modules iff $P(QA,t) = P(QB,t)$.

We are now prepared to prove the following result.

Theorem 3.1*: Let (A,d) be a differential cohomology Hopf algebra over the rational number field Q. Then

$$q_A : QHA \longrightarrow HQA$$

is an isomorphism.

Proof: Suppose that HA is of finite type. Consider the spectral sequence

$$E_r A \Longrightarrow HA$$

$$E_1 A = HE_0 A$$

introduced above . As (A,d) is a cohomology differential Hopf algebra over Q, A has a commutative multiplication and hence A is coprimitive. Hence E_0A is biprimitive.

Consider

$$d_0 : QE_0A \longrightarrow QE_0A.$$

One readily checks that

$$E_1A = S[HQE_0A]$$

where S[ ] denotes the symmetric , i.e., free commutative, algebra functor. (For this one uses the fact that $E_0 A \cong S[QE_0A]$ and elementary considerations.) Thus as

$$HQE_0A = HE_0^{1,*}A \cong s^1 HQ\ A$$

we find that

$$E_1 A \cong S[s^1 HQA]$$

as biprimitive differential Hopf algebras. Thus

$$PE_1 A = E_1^{1,*} A = Q E_1 A.$$

Now we assert that d_r , $r \geq 1$ is the first non-zero differential. As

$$E_1 A = E_r A$$

we find that under this identification

$$d_r(E_1^{1,*} A) = d_r(PE_r A) \subset PE_1 A = E_1^{1,*} A.$$

As d_r changes the filtration degree for $r \geq 1$ we find that we must have $d_r = 0$. Thus $d_r = 0$ for all $r \geq 1$ and hence $E_1 A = E_\infty A$ as claimed.

Thus we find that via its identification as an edge map in the spectral sequence $\{E_r, d_r\}$ the natural mapping

$$HA \longrightarrow HQA$$

is onto and hence

$$QHA \longrightarrow HQA$$

is onto.

Moreover as $E_\infty A = E_1 A$ is a free commutative algebra on the vector space $HQA = E_1^{1,*} A$, and HA is a free commutative algebra on the vector space QHA, the isomorphism of graded vector spaces

$$HA = E_1 A$$

entailed by the collapse of the spectral sequence yields the equality

$$P(QHA, t) = P(HQA, t)$$

of Poincaré series.

Therefore the epimorphism

$$QHA \longrightarrow HQA$$

is an isomorphism.

A direct limit argument yields the general case. □

There is an alternate way to view 3.1, which we describe now. Suppose that (A,d) is a differential homology Hopf algebra over Q. The natural composition

$$H(PA) \longrightarrow PHA \hookrightarrow HA$$

is then a map from a Lie algebra to the Lie algebra of a Hopf algebra. Thus there is induced a map of Hopf algebras

$$UHPA \longrightarrow HA$$

where U denotes the universal enveloping algebra functor. This map is then an isomorphism.

§4. Applications to Rational Homotopy

Let $(X, *)$ be a nicely pointed space, i.e., $* \varepsilon X$ is a neighbourhood deformation retraction in X. The rational homotopy of X is the connected graded Lie algebra over Q, denoted by $\mathcal{L}_*(X)$, defined by setting

$$\mathcal{L}_n(X) = \pi_{n+1}(X, *) \otimes_{\mathbb{Z}} Q$$

with Lie product induced by the Whitehead product on homotopy groups.

In this section we will reinterpret the results in the previous sections in terms of rational homotopy.

We begin by recalling that the Hurewicz map induces a natural morphism of degree -1

$$\mathcal{L}_*(X) \longrightarrow PH_*(X; Q)$$

which in the case that X is an H-space is an isomorphism by the results of Cartan-Serre [30; Appendix]. Combining the results of the previous section with the spectral sequence of Adams (1.2) we obtain:

<u>Theorem 4.1</u>: Let $(X, *)$ be a nicely pointed 1-connected space. Then there exists a natural second quadrant homology spectral sequence $\{E^r, d^r\}$ such that

$$E^r \Longrightarrow \mathcal{L}_*(X)$$

Moreover

$$E^1_{0,*} \cong P\,H_*(X;Q)$$

and the edge mapping

$$\mathcal{L}_*(X) \longrightarrow P\,H_*(X;Q)$$

coincides with the Hurewicz mapping.

<u>Proof</u>: We begin with the isomorphism of Adams (1.2)

$$H_*(\Omega X;\, Q) \cong H(\mathcal{F}C_*(X)).$$

From the cobar construction we obtain a spectral sequence

$$\{\hat{E}^r, \hat{d}^r\}$$

by filtering $\mathcal{F}C_*(X)$ as follows

$$F_{-p}\mathcal{F}C_*(X) = \sum_{n \geq p} C_*(X) \otimes \cdots \cdots \otimes C_*(X).$$

There results a spectral sequence with target $H_*(\Omega X;\, Q)$ and where

$$E^1 \cong \mathcal{F}H_*(X;Q) \cong T\,H_*(X;Q)$$

as differential Hopf algebras. Thus $\{\hat{E}^r, \hat{d}^r\}$ is a spectral sequence of homology Hopf algebras.

We next define a new spectral sequence $\{\check{E}^r, \check{d}^r\}$ by

$$\check{E}^r = P\hat{E}^r, \quad \check{d}^r = P\hat{d}^r .$$

It follows from the fact that $\hat{E}^r$ is a homology Hopf algebra <u>and</u> 3.1, that $\{\check{E}^r, \check{d}^r\}$ is indeed a spectral sequence. By the Cartan-Serre theorem the natural map

$$\mathcal{L}_*(X) \cong \pi_*(\Omega X, *) \longrightarrow P\,H_*(\Omega X;Q)$$

is an isomorphism. Thus we find that (1.2) induces a filtration on $\mathcal{L}_*(X)$ and

$$E^0\mathcal{L}_*(X) \cong P\hat{E}^\infty \equiv \check{E}^\infty .$$

Using the fact that

$$\check{E}^r_{0,*} = 0$$

we find that the reindexing

$$E^r_{p,q} = \check{E}^r_{p-1,\ q+1}$$

yields a spectral sequence

$$E^r \Longrightarrow \mathcal{L}_*(X).$$

To identify $E^1_{0,*}$ we observe that

$$PT(A) \cong A$$

for any A and use our above knowledge of $\hat{E}^1$. The assertion about the edge mapping is elementary. □

Remark: There is a sense in which the terms $E^1_{s,*}$ may be regarded as the sth-derived functor of the primitive functor P. However the language needed to make this precise would take us too far afield.

Acknowledgement: The result contained in 4.1 has also been discovered by A.K.Bousefield and D.L.Rector by completely different methods. Their approach seems closely related to the method employed by Rector [34] in his approach to the Eilenberg-Moore spectral sequence.

For the second application that we have in mind we shall need the following elementary technical result (see e.g.[14]):

Proposition 4.2: Let P denote the functor that assignes to a connected graded homology Hopf algebra over Q its associated connected graded Lie algebra of primitive elements. Then P preserves coproducts, i.e., if A, B are homology Hopf algebras then $P(A \amalg B)$ is a coproduct of PA and PB in the category of graded connected Lie algebras over Q. □

Applying this to (2.1) together with the sort of considerations employed in (4.1) we readily obtain:

Theorem 4.3: Let us suppose that

$$A \hookrightarrow X \longrightarrow B$$

is a cofibration sequence of 1-connected nicely pointed spaces. Then there is a natural spectral sequence of Lie algebras

$$E^r \Longrightarrow \mathcal{L}_*(X)$$

$$E^2 = \mathcal{L}_*(A) \amalg \mathcal{L}_*(B)$$

where $\amalg$ denotes the coproduct in the category of connected graded Lie algebras over Q. □

Remark: In some sense the above spectral sequence may be regarded as a Hilton-Eckmann dual to the Serre spectral sequence of a fibration. This duality is not complete and has been studied by G.E.Cooke (Princeton Thesis 1967).

We also obtain as a consequence the main result of [14] (although the proofs are almost identical).

Corollary 4.4: Let A and B be nicely pointed 1-connected spaces. Then the natural map

$$\mathcal{L}_*(A) \amalg \mathcal{L}_*(B) \longrightarrow \mathcal{L}_*(A \vee B)$$

is an isomorphism. □

References

1. J.F.Adams, On the Cobar Constructions, Proc. Nat. Acad. Sci. U.S.A. 42 (1956), 409-412.

2. J.F.Adams, On the Structure and Application of the Steenrod Algebra, Comment. Math. Helv. 32 (1958), 180-214.

3. J.F.Adams, On the Non-Existence of Elements of Hopf Invariant One, Ann. of Math. (2) 72 (1960), 20-104.

4. J.F.Adams, Lectures on Generalized Homology Theories, Springer Lecture Notes No.99 (1969).

5. D.W.Anderson and L.Hodgkin, The K-Theory of Eilenberg-MacLane Complexes, Topology 7 (1968).

6. M.F.Atiyah, Vector Bundles and the Künneth Formula, Topology 1 (1962), 245-248.

7. P.F.Baum, On the Cohomology of Homogeneous Spaces, Topology 7 (1968), 15-38.

8. J.Becker and R.J.Milgram, (to appear)

9. E.H.Brown, Twisted Tensor Products I, Ann. of Math. (2) 69 (1959), 223-246.

10. H.Cartan, Algebres d'Eilenberg-MacLane et Homotopie, Ecole Normale Superieure, Seminar H.Cartan 1954/1955.

11. H.Cartan and S.Eilenberg, Homological Algebra, Princeton University Press 1956.

12. H.Cartan and J.C.Moore, Périodicite des Groupes d'Homotopie Stables des Groupes Classiques, d'apres Bott, Ecole Normale Superieure, Seminar H.Cartan , 1959/60.

13. A.Clark, Homotopy Commutativity and the Moore Spectral Sequence, Pac. J. Math. 15 (1965), 65-74.

14. A.Clark and L.Smith, The Rational Homotopy of a Wedge, Pac. J. Math. 24 (1968), 241-246.

15. P.E.Conner and L.Smith, On the Complex Bordism of Finite Complexes, I.H.E.S. J. of Math. (to appear).

16. A.Dold, Halbexakte Homotopiefunktoren, Springer Lecture Notes No.12 (1966).

17. S.Eilenberg and J.C.Moore, Limits and Spectral Sequences, Topology 1 (1962), 1-24.

18. S.Eilenberg and J.C.Moore, Homology and Fibrations I, Comment. Math. Helv. 40 (1966), 199-236.

19. A.Heller, (to appear).

20. L.Hodgkin, An Equivariant Künneth Formula in K-Theory, Warwick University Preprint (196?).

21. L.Hodgkin, Notes Towards a Geometric Eilenberg-Moore Spectral Sequence, Preprint (1968).

22. S-T.Hu, Homotopy Theory, Academic Press, New York, 1959.

23. I.M.James, Ex-Homotopy Theory, Ill. J. of Math. (to appear).

24. D.M.Kan, Adjoint Functors, Trans. of A.M.S. 87 (1958) 294-329.

25. L.Kristensen, On the Cohomology of Spaces with Two Non-Vanishing Homotopy Groups, Math. Scand. 12 (1963), 83-105.

26. W.S.Massey, Exact Couples in Algebraic Topology I-V, Ann. of Math. 56 (1952),363-396, 57(1953) 248-280.

27. W.S.Massey and F.P.Peterson, The Cohomology Structure of Certain Fibre Spaces I, Topology 4 (1965), 47-65.

28. J.P.Meyer, Relative Stable Homotopy Theory, (to appear).

29. J.W.Milnor, On Spaces Having the Homotopy Type of a CW-Complex, Trans. of A.M.S. 90 (1959), 272-280.

30. J.W.Milnor and J.C.Moore, On the Structure of Hopf Algebras, Ann. of Math.(2) 81 (1965), 211-264.

31. J.C.Moore and L.Smith, Hopf Algebras and Multiplicative Fibrations I,II, Am. J. of Math. 90 (1968), 752-780,1113-1150.

32. S.MacLane, Homology, Springer-Verlag/ Academic Press, Heidelberg / New York, 1962.

33. J.F.McClendon, (to appear).

34. D.L.Rector, (to appear).

35. J.P.Serre, Homologie Singuliere des Espace Fibré, Applications, Ann. of Math. (2) 54, 425-505.

36. J.P.Serre, Cohomologie Modulo 2 des Complexes d'Eilenberg-MacLane, Comment. Math. Helv. 27 (1954), 198-232.

37. W.S.Singer, Connective Fiberings Over BU and U, Topology 7 (1968), 189-225.

38. L.Smith, Cohomology of Two Stage Postnikov Systems, Ill. J. of Math. 11 (1967), 310-329.

39. L.Smith, Homological Algebra and the Eilenberg-Moore Spectral Sequence, Trans. of A.M.S. 129 (1967), 58-93.

40. L.Smith, Primitive Loop Spaces, Topology 7 (1968),121-124.

41. L.Smith, On the Relation Between Spherical and Primitive Homology Classes in Topological Groups, Topology 8 (1969), 69-80.

42. L.Smith, Split Extensions of Hopf Algebras and Semi-Tensor Products, Math. Scand. 23 (1969)

43. L.Smith, On the Construction of the Eilenberg-Moore Spectral Sequence, Bull. A.M.S. (1969).

44. L.Smith, Hopf Fibration Towers and the Unstable Adams Spectral Sequence, Proceedings of the Conference on Applications of Categorical Algebra (to appear).

45. L.Smith, On the Künneth Theorem I,II (to appear).

46. L.Smith, On the Differentials in the Eilenberg-Moore Spectral Sequence, (to appear).

47. E.H.Spanier, Algebraic Topology, McGraw Hill, New York 1966.

48. N.E.Steenrod, A Convenient Category of Topological Spaces, Mich. J. Math. 14 (1967), 133-152.

49. N.E.Steenrod and D.B.A.Epstein, Cohomology Operations, Ann. of Math. Studies No.50.

50. G.W.Whitehead, Generalized Homology Theories, Trans. of A.M.S. 102 (1962), 227-283.

Offsetdruck: Julius Beltz, Weinheim/Bergstr.

Lecture Notes in Mathematics

Bisher erschienen/Already published

Vol. 1: J. Wermer, Seminar über Funktionen-Algebren. IV, 30 Seiten. 1964. DM 3,80 / $ 1.10

Vol. 2: A. Borel, Cohomologie des espaces localement compacts d'après. J. Leray. IV, 93 pages. 1964. DM 9,– / $ 2.60

Vol. 3: J. F. Adams, Stable Homotopy Theory. Third edition. IV, 78 pages. 1969. DM 8,– / $ 2.20

Vol. 4: M. Arkowitz and C. R. Curjel, Groups of Homotopy Classes. 2nd. revised edition. IV, 36 pages. 1967. DM 4,80 / $ 1.40

Vol. 5: J.-P. Serre, Cohomologie Galoisienne Troisième édition. VIII, 214 pages. 1965. DM 18,– / $ 5.00

Vol. 6: H. Hermes, Eine Termlogik mit Auswahloperator. IV, 42 Seiten. 1965. DM 5,80 / $ 1.60

Vol. 7: Ph. Tondeur, Introduction to Lie Groups and Transformation Groups. Second edition. VIII, 176 pages. 1969. DM 14,– / $ 3.80

Vol. 8: G. Fichera, Linear Elliptic Differential Systems and Eigenvalue Problems. IV, 176 pages. 1965. DM 13,50 / $ 3.80

Vol. 9: P. L. Ivănescu, Pseudo-Boolean Programming and Applications. IV, 50 pages. 1965. DM 4,80 / $ 1.40

Vol. 10: H. Lüneburg, Die Suzukigruppen und ihre Geometrien. VI, 111 Seiten. 1965. DM 8,– / $ 2.20

Vol. 11: J.-P. Serre, Algèbre Locale. Multiplicités. Rédigé par P. Gabriel. Seconde édition. VIII, 192 pages. 1965. DM 12,– / $ 3.30

Vol. 12: A. Dold, Halbexakte Homotopiefunktoren. II, 157 Seiten. 1966. DM 12,– / $ 3.30

Vol. 13: E. Thomas, Seminar on Fiber Spaces. IV, 45 pages. 1966. DM 4,80 / $ 1.40

Vol. 14: H. Werner, Vorlesung über Approximationstheorie. IV, 184 Seiten und 12 Seiten Anhang. 1966. DM 14,– / $ 3.90

Vol. 15: F. Oort, Commutative Group Schemes. VI, 133 pages. 1966. DM 9,80 / $ 2.70

Vol. 16: J. Pfanzagl and W. Pierlo, Compact Systems of Sets. IV, 48 pages. 1966. DM 5,80 / $ 1.60

Vol. 17: C. Müller, Spherical Harmonics. IV, 46 pages. 1966. DM 5,– / $ 1.40

Vol. 18: H.-B. Brinkmann und D. Puppe, Kategorien und Funktoren. XII, 107 Seiten. 1966. DM 8,– / $ 2.20

Vol. 19: G. Stolzenberg, Volumes, Limits and Extensions of Analytic Varieties. IV, 45 pages. 1966. DM 5,40 / $ 1.50

Vol. 20: R. Hartshorne, Residues and Duality. VIII, 423 pages. 1966. DM 20,– / $ 5.50

Vol. 21: Seminar on Complex Multiplication. By A. Borel, S. Chowla, C. S. Herz, K. Iwasawa, J.-P. Serre. IV, 102 pages. 1966. DM 8,– /$ 2.20

Vol. 22: H. Bauer, Harmonische Räume und ihre Potentialtheorie. IV, 175 Seiten. 1966. DM 14,– / $ 3.90

Vol. 23: P. L. Ivănescu and S. Rudeanu, Pseudo-Boolean Methods for Bivalent Programming. 120 pages. 1966. DM 10,– / $ 2.80

Vol. 24: J. Lambek, Completions of Categories. IV, 69 pages. 1966. DM 6,80 / $ 1.90

Vol. 25: R. Narasimhan, Introduction to the Theory of Analytic Spaces. IV, 143 pages. 1966. DM 10,– / $ 2.80

Vol. 26: P.-A. Meyer, Processus de Markov. IV, 190 pages. 1967. DM 15,– / $ 4.20

Vol. 27: H. P. Künzi und S. T. Tan, Lineare Optimierung großer Systeme. VI, 121 Seiten. 1966. DM 12,– / $ 3.30

Vol. 28: P. E. Conner and E. E. Floyd, The Relation of Cobordism to K-Theories. VIII, 112 pages. 1966. DM 9,80 / $ 2.70

Vol. 29: K. Chandrasekharan, Einführung in die Analytische Zahlentheorie. VI, 199 Seiten. 1966. DM 16,80 / $ 4.70

Vol. 30: A. Frölicher and W. Bucher, Calculus in Vector Spaces without Norm. X, 146 pages. 1966. DM 12,– / $ 3.30

Vol. 31: Symposium on Probability Methods in Analysis. Chairman. D. A. Kappos.IV, 329 pages. 1967. DM 20,– / $ 5.50

Vol. 32: M. André, Méthode Simpliciale en Algèbre Homologique et Algèbre Commutative. IV, 122 pages. 1967. DM 12,– / $ 3.30

Vol. 33: G. I. Targonski, Seminar on Functional Operators and Equations. IV, 110 pages. 1967. DM 10,– / $ 2.80

Vol. 34: G. E. Bredon, Equivariant Cohomology Theories. VI, 64 pages. 1967. DM 6,80 / $ 1.90

Vol. 35: N. P. Bhatia and G. P. Szegö, Dynamical Systems. Stability Theory and Applications. VI, 416 pages. 1967. DM 24,– / $ 6.60

Vol. 36: A. Borel, Topics in the Homology Theory of Fibre Bundles. VI, 95 pages. 1967. DM 9,– / $ 2.50

Vol. 37: R. B. Jensen, Modelle der Mengenlehre. X, 176 Seiten. 1967. DM 14,– / $ 3.90

Vol. 38: R. Berger, R. Kiehl, E. Kunz und H.-J. Nastold, Differentialrechnung in der analytischen Geometrie IV, 134 Seiten. 1967 DM 12,– / $ 3.30

Vol. 39: Séminaire de Probabilités I. II, 189 pages. 1967. DM 14,– / $ 3.90

Vol. 40: J. Tits, Tabellen zu den einfachen Lie Gruppen und ihren Darstellungen. VI, 53 Seiten. 1967. DM 6,80 / $ 1.90

Vol. 41: A. Grothendieck, Local Cohomology. VI, 106 pages. 1967. DM 10,– / $ 2.80

Vol. 42: J. F. Berglund and K. H. Hofmann, Compact Semitopological Semigroups and Weakly Almost Periodic Functions. VI, 160 pages. 1967. DM 12,– / $ 3.30

Vol. 43: D. G. Quillen, Homotopical Algebra. VI, 157 pages. 1967. DM 14,– / $ 3.90

Vol. 44: K. Urbanik, Lectures on Prediction Theory. IV, 50 pages. 1967. DM 5,80 / $ 1.60

Vol. 45: A. Wilansky, Topics in Functional Analysis. VI, 102 pages. 1967. DM 9,60 / $ 2.70

Vol. 46: P. E. Conner, Seminar on Periodic Maps. IV, 116 pages. 1967. DM 10,60 / $ 3.00

Vol. 47: Reports of the Midwest Category Seminar I. IV, 181 pages. 1967. DM 14,80 / $ 4.10

Vol. 48: G. de Rham, S. Maumary et M. A. Kervaire, Torsion et Type Simple d'Homotopie. IV, 101 pages. 1967. DM 9,60 / $ 2.70

Vol. 49: C. Faith, Lectures on Injective Modules and Quotient Rings. XVI, 140 pages. 1967. DM 12,80 / $ 3.60

Vol. 50: L. Zalcman, Analytic Capacity and Rational Approximation. VI, 155 pages. 1968. DM 13.20 / $ 3.70

Vol. 51: Séminaire de Probabilités II. IV, 199 pages. 1968. DM 14,– / $ 3.90

Vol. 52: D. J. Simms, Lie Groups and Quantum Mechanics. IV, 90 pages. 1968. DM 8,– / $ 2.20

Vol. 53: J. Cerf, Sur les difféomorphismes de la sphère de dimension trois (Γ_4= O). XII, 133 pages. 1968. DM 12,– / $ 3.30

Vol. 54: G. Shimura, Automorphic Functions and Number Theory. VI, 69 pages. 1968. DM 8,– / $ 2.20

Vol. 55: D. Gromoll, W. Klingenberg und W. Meyer, Riemannsche Geometrie im Großen. VI, 287 Seiten. 1968. DM 20,– / $ 5.50

Vol. 56: K. Floret und J. Wloka, Einführung in die Theorie der lokalkonvexen Räume. VIII, 194 Seiten. 1968. DM 16,– / $ 4.40

Vol. 57: F. Hirzebruch und K. H. Mayer, O (n)-Mannigfaltigkeiten, exotische Sphären und Singularitäten. IV, 132 Seiten. 1968. DM 10,80/ $ 3.00

Vol. 58: Kuramochi Boundaries of Riemann Surfaces. IV, 102 pages. 1968. DM 9,60 / $ 2.70

Vol. 59: K. Jänich, Differenzierbare G-Mannigfaltigkeiten. VI, 89 Seiten. 1968. DM 8,– / $ 2.20

Vol. 60: Seminar on Differential Equations and Dynamical Systems. Edited by G. S. Jones. VI, 106 pages. 1968. DM 9,60 / $ 2.70

Vol. 61: Reports of the Midwest Category Seminar II. IV, 91 pages. 1968. DM 9,60 / $ 2.70

Vol. 62: Harish-Chandra, Automorphic Forms on Semisimple Lie Groups X, 138 pages. 1968. DM 14,– / $ 3.90

Vol. 63: F. Albrecht, Topics in Control Theory. IV, 65 pages. 1968. DM 6,80 / $ 1.90

Vol. 64: H. Berens, Interpolationsmethoden zur Behandlung von Approximationsprozessen auf Banachräumen. VI, 90 Seiten. 1968. DM 8,– / $ 2.20

Vol. 65: D. Kölzow, Differentiation von Maßen. XII, 102 Seiten. 1968. DM 8,– / $ 2.20

Vol. 66: D. Ferus, Totale Absolutkrümmung in Differentialgeometrie und -topologie. VI, 85 Seiten. 1968. DM 8,– / $ 2.20

Vol. 67: F. Kamber and P. Tondeur, Flat Manifolds. IV, 53 pages. 1968. DM 5,80 / $ 1.60

Vol. 68: N. Boboc et P. Mustată, Espaces harmoniques associés aux opérateurs différentiels linéaires du second ordre de type elliptique. VI, 95 pages. 1968. DM 8,60 / $ 2.40

Vol. 69: Seminar über Potentialtheorie. Herausgegeben von H. Bauer. VI, 180 Seiten. 1968. DM 14,80 / $ 4.10

Vol. 70: Proceedings of the Summer School in Logic. Edited by M. H. Löb. IV, 331 pages. 1968. DM 20,– / $ 5.50

Vol. 71: Séminaire Pierre Lelong (Analyse), Année 1967 – 1968. VI, 19 pages. 1968. DM 14,– / $ 3.90

Bitte wenden / Continued

Vol. 72: The Syntax and Semantics of Infinitary Languages. Edited by J. Barwise. IV, 268 pages. 1968. DM 18,– / $ 5.00

Vol. 73: P. E. Conner, Lectures on the Action of a Finite Group. IV, 123 pages. 1968. DM 10,– / $ 2.80

Vol. 74: A. Fröhlich, Formal Groups. IV, 140 pages. 1968. DM 12,– / $ 3.30

Vol. 75: G. Lumer, Algèbres de fonctions et espaces de Hardy. VI, 80 pages. 1968. DM 8,– / $ 2.20

Vol. 76: R. G. Swan, Algebraic K-Theory. IV, 262 pages. 1968. DM 18,– / $ 5.00

Vol. 77: P.-A. Meyer, Processus de Markov: la frontière de Martin. IV, 123 pages. 1968. DM 10,– / $ 2.80

Vol. 78: H. Herrlich, Topologische Reflexionen und Coreflexionen. XVI, 166 Seiten. 1968. DM 12,– / $ 3.30

Vol. 79: A. Grothendieck, Catégories Cofibrées Additives et Complexe Cotangent Relatif. IV, 167 pages. 1968. DM 12,– / $ 3.30

Vol. 80: Seminar on Triples and Categorical Homology Theory. Edited by B. Eckmann. IV, 398 pages. 1969. DM 20,– / $ 5.50

Vol. 81: J.-P. Eckmann et M. Guenin, Méthodes Algébriques en Mécanique Statistique. VI, 131 pages. 1969. DM 12,– / $ 3.30

Vol. 82: J. Wloka, Grundräume und verallgemeinerte Funktionen. VIII, 131 Seiten. 1969. DM 12,– / $ 3.30

Vol. 83: O. Zariski, An Introduction to the Theory of Algebraic Surfaces. IV, 100 pages. 1969. DM 8,– / $ 2.20

Vol. 84: H. Lüneburg, Transitive Erweiterungen endlicher Permutationsgruppen. IV, 119 Seiten. 1969. DM 10,– / $ 2.80

Vol. 85: P. Cartier et D. Foata, Problèmes combinatoires de commutation et réarrangements. IV, 88 pages. 1969. DM 8,– / $ 2.20

Vol. 86: Category Theory, Homology Theory and their Applications I. Edited by P. Hilton. VI, 216 pages. 1969. DM 16,– / $ 4.40

Vol. 87: M. Tierney, Categorical Constructions in Stable Homotopy Theory. IV, 65 pages. 1969. DM 6,– / $ 1.70

Vol. 88: Séminaire de Probabilités III. IV, 229 pages. 1969. DM 18,– / $ 5.00

Vol. 89: Probability and Information Theory. Edited by M. Behara, K. Krickeberg and J. Wolfowitz. IV, 256 pages. 1969. DM 18,– / $ 5.00

Vol. 90: N. P. Bhatia and O. Hajek, Local Semi-Dynamical Systems. II, 157 pages. 1969. DM 14,– / $ 3.90

Vol. 91: N. N. Janenko, Die Zwischenschrittmethode zur Lösung mehrdimensionaler Probleme der mathematischen Physik. VIII, 194 Seiten. 1969. DM 16,80 / $ 4.70

Vol. 92: Category Theory, Homology Theory and their Applications II. Edited by P. Hilton. V, 308 pages. 1969. DM 20,– / $ 5.50

Vol. 93: K. R. Parthasarathy, Multipliers on Locally Compact Groups. III, 54 pages. 1969. DM 5,60 / $ 1.60

Vol. 94: M. Machover and J. Hirschfeld, Lectures on Non-Standard Analysis. VI, 79 pages. 1969. DM 6,– / $ 1.70

Vol. 95: A. S. Troelstra, Principles of Intuitionism. II, 111 pages. 1969. DM 10,– / $ 2.80

Vol. 96: H.-B. Brinkmann und D. Puppe, Abelsche und exakte Kategorien, Korrespondenzen. V, 141 Seiten. 1969. DM 10,– / $ 2.80

Vol. 97: S. O. Chase and M. E. Sweedler, Hopf Algebras and Galois theory. II, 133 pages. 1969. DM 10,– / $ 2.80

Vol. 98: M. Heins, Hardy Classes on Riemann Surfaces. III, 106 pages. 1969. DM 10,– / $ 2.80

Vol. 99: Category Theory, Homology Theory and their Applications III. Edited by P. Hilton. IV, 489 pages. 1969. DM 24,– / $ 6.60

Vol. 100: M. Artin and B. Mazur, Etale Homotopy. II, 196 Seiten. 1969. DM 12,– / $ 3.30

Vol. 101: G. P. Szegö et G. Treccani, Semigruppi di Trasformazioni Multivoche. VI, 177 pages. 1969. DM 14,– / $ 3.90

Vol. 102: F. Stummel, Rand- und Eigenwertaufgaben in Sobolewschen Räumen. VIII, 386 Seiten. 1969. DM 20,– / $ 5.50

Vol. 103: Lectures in Modern Analysis and Applications I. Edited by C. T. Taam. VII, 162 pages. 1969. DM 12,– / $ 3.30

Vol. 104: G. H. Pimbley, Jr., Eigenfunction Branches of Nonlinear Operators and their Bifurcations. II, 128 pages. 1969. DM 10,– / $ 2.80

Vol. 105: R. Larsen, The Multiplier Problem. VII, 284 pages. 1969. DM 18,– / $ 5.00

Vol. 106: Reports of the Midwest Category Seminar III. Edited by S. Mac Lane. III, 247 pages. 1969. DM 16,– / $ 4.40

Vol. 107: A. Peyerimhoff, Lectures on Summability. III, 111 pages. 1969. DM 8,– / $ 2.20

Vol. 108: Algebraic K-Theory and its Geometric Applications. Edited by R. M.F. Moss and C. B. Thomas. IV, 86 pages. 1969. DM 6,– / $ 1.70

Vol. 109: Conference on the Numerical Solution of Differential Equations. Edited by J. Ll. Morris. VI, 275 pages. 1969. DM 18,– / $ 5.00

Vol. 110: The Many Facets of Graph Theory. Edited by G. Chartrand and S. F. Kapoor. VIII, 290 pages. 1969. DM 18,– / $ 5.00

Vol. 111: K. H. Mayer, Relationen zwischen charakteristischen Zahlen. III, 99 Seiten. 1969. DM 8,– / $ 2.20

Vol. 112: Colloquium on Methods of Optimization. Edited by N. N. Moiseev. IV, 293 pages. 1970. DM 18,– / $ 5.00

Vol. 113: R. Wille, Kongruenzklassengeometrien. III, 99 Seiten. 1970. DM 8,– / $ 2.20

Vol. 114: H. Jacquet and R. P. Langlands, Automorphic Forms on GL (2). VII, 548 pages. 1970. DM 24,– / $ 6.60

Vol. 115: K. H. Roggenkamp and V. Huber-Dyson, Lattices over Orders I. XIX, 290 pages. 1970. DM 18,– / $ 5.00

Vol. 116: Séminaire Pierre Lelong (Analyse) Année 1969. IV, 195 pages. 1970. DM 14,– / $ 3.90

Vol. 117: Y. Meyer, Nombres de Pisot, Nombres de Salem et Analyse Harmonique. 63 pages. 1970. DM 6,– / $ 1.70

Vol. 118: Proceedings of the 15th Scandinavian Congress, Oslo 1968. Edited by K. E. Aubert and W. Ljunggren. IV, 162 pages. 1970. DM 12,– / $ 3.30

Vol. 119: M. Raynaud, Faisceaux amples sur les schémas en groupes et les espaces homogènes. III, 219 pages. 1970. DM 14,– / $ 3.90

Vol. 120: D. Siefkes, Büchi's Monadic Second Order Successor Arithmetic. XII, 130 Seiten. 1970. DM 12,– / $ 3.30

Vol. 121: H. S. Bear, Lectures on Gleason Parts. III, 47 pages. 1970. DM 6,– / $ 1.70

Vol. 122: H. Zieschang, E. Vogt und H.-D. Coldewey, Flächen und ebene diskontinuierliche Gruppen. VIII, 203 Seiten. 1970. DM 16,– / $ 4.40

Vol. 123: A. V. Jategaonkar, Left Principal Ideal Rings. VI, 145 pages. 1970. DM 12,– / $ 3.30

Vol. 124: Séminare de Probabilités IV. Edited by P. A. Meyer. IV, 282 pages. 1970. DM 20,– / $ 5.50

Vol. 125: Symposium on Automatic Demonstration. V, 310 pages. 1970. DM 20,– / $ 5.50

Vol. 126: P. Schapira, Théorie des Hyperfonctions. XI, 157 pages. 1970. DM 14,– / $ 3.90

Vol. 127: I. Stewart, Lie Algebras. IV, 97 pages. 1970. DM 10,– / $ 2.80

Vol. 128: M. Takesaki, Tomita's Theory of Modular Hilbert Algebras and its Applications. II, 123 pages. 1970. DM 10,– / $ 2.80

Vol. 129: K. H. Hofmann, The Duality of Compact Semigroups and C*-Bigebras. XII, 142 pages. 1970. DM 14,– / $ 3.90

Vol. 130: F. Lorenz, Quadratische Formen über Körpern. II, 77 Seiten. 1970. DM 8,– / $ 2.20

Vol. 131: A Borel et al., Seminar on Algebraic Groups and Related Finite Groups. VII, 321 pages. 1970. DM 22,– / $ 6.10

Vol. 132: Symposium on Optimization. III, 348 pages. 1970. DM 22,– / $ 6.10

Vol. 133: F. Topsøe, Topology and Measure. XIV, 79 pages. 1970. DM 8,– / $ 2.20

Vol. 134: L. Smith, Lectures on the Eilenberg-Moore Spectral Sequence. VII, 142 pages. 1970. DM 14,– / $ 3.90